# Flow Cytometry

## First Principles

R

# Flow Cytometry

## First Principles

**Alice Longobardi Givan, PhD**

The Norris Cotton Cancer Center
Flow Cytometry Laboratory
and Department of Physiology
Dartmouth-Hitchcock Medical Center
Lebanon, New Hampshire

*formerly of*

Department of Surgery
The Medical School
University of Newcastle upon Tyne
Newcastle upon Tyne, England



WILEY-LISS

A JOHN WILEY & SONS, INC., PUBLICATION
New York • Chichester • Brisbane • Toronto • Singapore

**Address All Inquiries to the Publisher**
**Wiley-Liss, Inc., 605 Third Avenue, New York, NY 10158-0012**



**Printed in the United States of America.**



**Library of Congress Cataloging-in-Publication Data**

Givan, Alice Longobardi.
Flow cytometry : first principles / Alice Longobardi Givan.
p. cm.
Includes bibliographical references and index.
ISBN 0-471-56095-2
1. Flow cytometry. I. Title.
[DNLM: 1. Flow Cytometry—methods. QH 585.5.F56 G539f]
QH585.5.F56G58 1992
574.87'028—dc20
DNLM/DLC
for Library of Congress 92-5004
CIP

**The text of this book is printed on acid-free paper.**

10 9 8 7 6 5 4

*This book is dedicated to my parents,*
*Violet Litwin Longobardi and Vincent Longobardi, Jr.,*
*with gratitude for the example they set,*
*with pride in their achievements,*
*and with love*

# Contents

# Preface

Although flow cytometry is simply a technique that is useful in certain fields of scientific endeavor, there is, at the same time, something special about it. Few other techniques involve specialists from so many different backgrounds. Anyone working with flow systems for any length of time will realize that computer buffs, electronics experts, mathematicians, laser technologists, and organic chemists rub shoulders with biologists, physicians, and surgeons around the flow cytometer bench.

And it is not just a casual rubbing of shoulders, in passing, so to speak. Many of the specialists involved in flow cytometry might, if asked, call themselves *flow cytometrists*, because the second aspect of flow cytometry that distinguishes it from many other techniques is that flow cytometry has itself become a "field." Indeed, it is a field of endeavor and of expertise that has captured the imaginations of many people. And it is a field to which growing numbers of people are attracted. As a result of this attraction, there exists a spirit of camaraderie; flow cytometry societies, clubs, meetings, journals, courses, and books abound.

A third aspect of flow cytometry (known sometimes simply as *cytometry* or by the acronym for fluorescence-activated cell sorter, *FACS*, or even more familiarly as *flow*) that distinguishes it from many other techniques is the way in which its wide and increasing usefulness has continued to surprise even those who consider themselves experts. What began as a clever technique for looking at a very limited range of problems is now being used in universities, in hospitals, within industry, at marine stations, and on board ships; plans exist for future use on board space ships as well. The applications of flow cytometry are proliferating rapidly, both in the direction of theoretical science, with botany, molecular biology, embryology, marine ecology, genetics, microbiology, and immunology, for example, all represented; and in the direction of clinical diagnosis and medical practice, with hematology, dermatology, bacteriology, pathology, oncology, obstetrics, surgery, genetics, and immunology becoming involved. We are, at present, living through what appears to be a rapid phase in flow cytometry's growth curve (see Fig. 1).

Because flow cytometry is an unusual field, bringing together people with differing scientific backgrounds at meetings, on editorial boards, in hospital wards, and at laboratory benches, and reaching increasing numbers of workers in new and unpredicted areas of endeavor, there is, as a result, a need to provide both recent and potential entrants into this diverse community with a common basis of knowledge—so that we can all understand the vocabulary, the assumptions, the strengths, and the weaknesses of the technology involved. I have for several years given a course for new and future users of the flow cytometers at Newcastle University. The course has attempted to present enough technical background to enable students, scientists, technologists, and clinicians to read the literature critically, to evaluate the benefits of the technique realistically, and, if tempted, to design effective protocols and interpret the results.

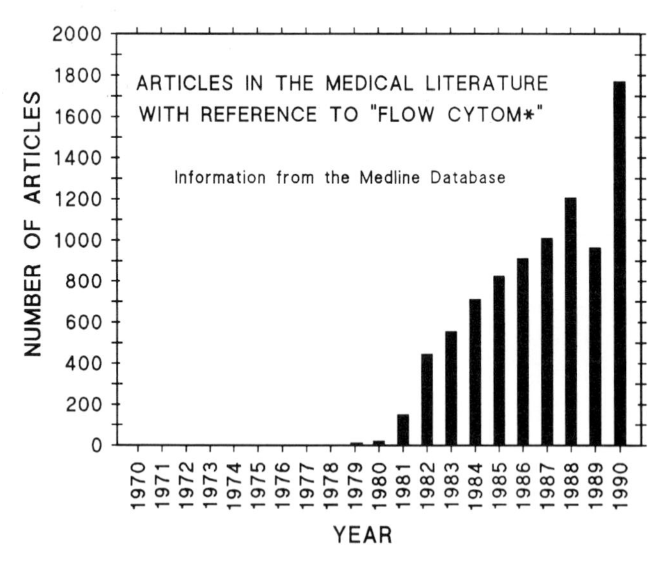


**Fig. 1.** Increasing reference to flow cytometry in the medical literature over the past two decades.

I have tried to describe the theory of flow cytometry in a way that also provides a firm (and accurate) foundation for those few who will want to study the technique in greater depth. Details of methodology have been avoided, but the course has attempted to give enough information about applications to provide concrete examples of general concepts and to allow some appreciation of the range of practical goals that the instruments are able to achieve. The course is presented as a series of five 2-hour lectures; with some expansion (but with little change in style or objectives), those lectures are the basis for this book.

# Acknowledgments

Realizing how much I have learned and continue to learn from others, I hesitate to single out a few names for particular mention. However, with the disclaimer that any list of people to whom I am indebted is not meant to be and, indeed, could never be complete, I must thank here the following friends, mentors, and colleagues who have had a very direct impact on the writing of this book: George Proud, for having had enough insight into the importance of a flow cytometer for transplantation surgery to want to have one; Ivan Johnston and Ross Taylor, for hospitality (and the Harker Bequest Fund, the Northern Counties Kidney Research Fund and the Newcastle Health Authority, for financial assistance) within the Department of Surgery at Newcastle University; Brian Shenton, for introducing me to the field of flow cytometry and to the joys and hazards of clinical research; Mike White, for bailing me out (often figuratively and once or twice literally) on so many occasions; Paul Dunnigan, for teaching me about lasers (and also about fuses, relays, and loose wires); Ian Brotherick, for the animal amplification figure, and both he and Alison Mitcheson for good humor in the lab beyond all reasonable expectation; Terry Godley, for being an extremely good and (more importantly) a very communicative flow cytometry service engineer; Ray Joyce, for considerable assistance with the design and drawing of many of the diagrams; the scientists and clinicians at Newcastle University and Durham University (many acknowledged in the figure legends), for providing me with a continuous source of interesting and well-prepared cell material on which to practice my flow technique; Paul Guyre, for giving me a supportive and remarkably enjoyable re-introduction to science in the New World; Daryll Green, for giving me the benefit of his expertise in both physics and flow cytometry by reading and commenting about the chapters on instrumentation and information; Brian Crawford, for being a literate and encouraging editor in the face of my rank inexperience; the late Robert L. Conner, for providing me with that critical first of my still-continuing scientific apprenticeships; Curt Givan, for his unfailing loyalty and for his skill in reading the entire manuscript with two eyes—one eye that of an old-fashioned grammarian who abhors dangling participles and the other that of a modern scientist who knows nothing about flow cytometry; Ben Givan for the two drawings in Figure 8.1, Becky Givan, for organizational magic when it was very badly needed, and both kids for *lots* of encouragement and some pretty funny suggestions for a title.

A.L.G.

# 1

# The Past as Prologue

Flow cytometry, like most scientific developments, has roots firmly grounded in history. In particular, flow technology finds its intellectual antecedents in microscopy, in blood cell counting instruments, and in the ink jet technology that was, in the 1960s, being developed for computer printers. It was the coming together of these three strands of endeavor that provided the basis for the development of the first flow cytometers. Because thorough accounts of the history of flow cytometry have been written elsewhere (and make a fascinating story for those interested in the history of science), I cover past history here in just enough detail to give readers a perspective as to why current instruments have developed as they have.

Microscopes have, since the seventeenth century, been used to examine cells and tissue sections. Particularly since the end of the nineteenth century, stains have been developed that make various cellular constituents visible; in the 1940s and 1950s, fluorescence microscopy began to be used in conjunction with fluorescent stains for nucleic acids in order to detect malignant cells. With the advent of antibody technology and the work of Coons in linking antibodies with fluorescent tags, the use of fluorescent stains gained wider and more specific applications. In particular, cell suspensions or tissue sections are now routinely stained with antibodies specific for antigenic markers of cell type. The antibodies are either directly or indirectly conjugated to fluorescent compounds (most usually fluorescein or rhodamine). And the cellular material can then be examined on a glass slide under a microscope fitted with appropriate lamp and filters so that the fluorescence of the cells can be excited and observed (Fig. 1.1). The fluorescence microscope allows the user to see the cells, to identify them both in terms of their structural patterns and their orientation within tissues,

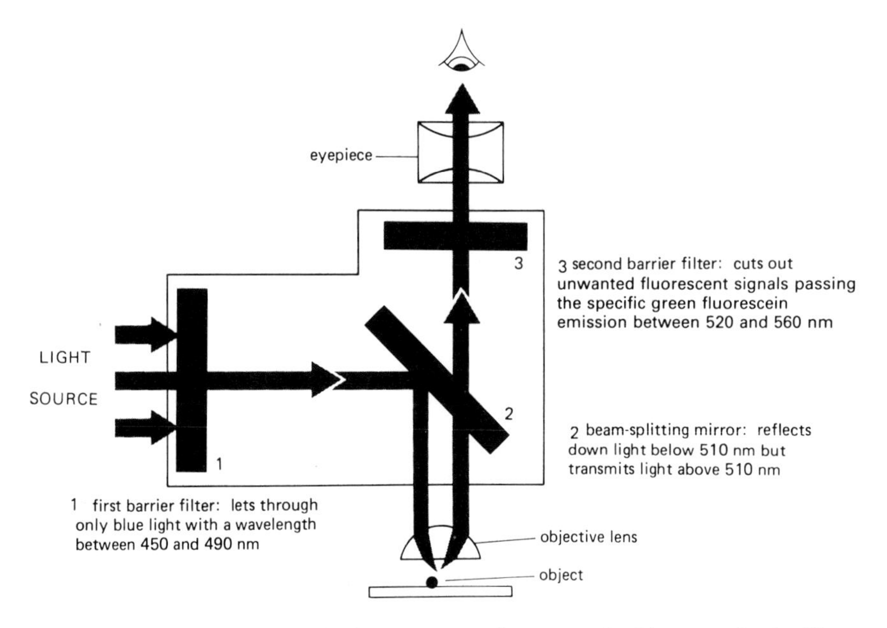


**Fig. 1.1.** The optical system of a fluorescence microscope. In this example, the filters and mirrors are set for detection of fluorescein fluorescence. From Alberts et al., (1989).

and then to determine whether and in what pattern they fluoresce when stained with one or another of the specific stains available. In addition, a microscope can also be fitted with a photodetector, which will then record the intensity of fluorescence arising from the field in view. The logical extension of this technique is image analysis cytometry, a recent development in microscopy that now allows precise quantitation of fluorescence intensity patterns in detail within that field of view. The development of monoclonal antibody technology (for which Kohler and Milstein were awarded the Nobel Prize) has led to a vast increase in the number of cellular components that can be specifically stained and that can be used to classify cells. Whereas monoclonal antibody techniques are not directly related to the development of flow technology, their invention was a serendipitous event that had great impact on the potential utility of flow systems.

In 1934, Andrew Moldavan in Montreal took a first step from static microscopy toward a flowing system. He suggested the development of an apparatus to count red blood cells and neutral-red-stained yeast cells as they were forced through a capillary on a microscope stage. A photodetector attached to the microscope eyepiece would register each passing cell. Although it is unclear from Moldavan's paper whether he actually ever built this cytometer, the development of staining procedures over the next 30 years made it obvious that the technique he suggested could be useful not simply for counting the number of cells but also for quantitating their characteristics.

In the mid-1960s, as a result of a desire to automate cervical cytology screening, Louis Kamentsky in New York developed a microscope-based spectrophotometer (on the pattern of the one suggested by Moldavan) that measured and recorded UV absorption and the scatter of blue light ("as an alternative to mimicking the complex scanning methods of the human microscopist") from cells flowing "at rates exceeding 500 cells per second" past a microscope objective. And then, in 1967, Kamentsky and Melamed elaborated this design into a sorting flow cell (Fig. 1.2) that provided for the electronic actuation of a syringe to pull cells with high absorption/scatter ratios out of the flow stream. These "suspicious" cells could then be subjected to detailed microscopic analysis. In 1969, Dittrich and Göhde in Germany described a flow chamber for a microscope-based system whereby fluorescence intensity histograms could be generated based on the ethidium bromide fluorescence of alcohol-fixed cells.

Even before these developments in microscope technology, so-called Coulter technology had been developed for analysis of blood cells. In the 1950s instruments were produced that counted cells as they flowed in a liquid stream; analysis was based on the amount by which the cells increased the electrical resistance of an orifice as they displaced isotonic saline solution while flowing through it. Cells were thereby classified more or less on the basis of their volume, since larger cells had greater electrical resistance. These Coulter counters soon became essential equipment in hospital hematology laboratories, allowing the rapid and automated counting of white and red blood cells. They actually incorporated many of the features of analysis that we now think of as being typical of flow cytometry: the rapid flow of single cells in file through an orifice, the detection of electrical signals from those cells, and the automated analysis of those signals.

At the same time as Kamentsky's work on cervical screening, Mack Fulwyler at the Los Alamos laboratories had decided to investigate a problem well known to everyone looking at red blood cells in Coulter counters. Red cells were known to show a bimodal distribution of their

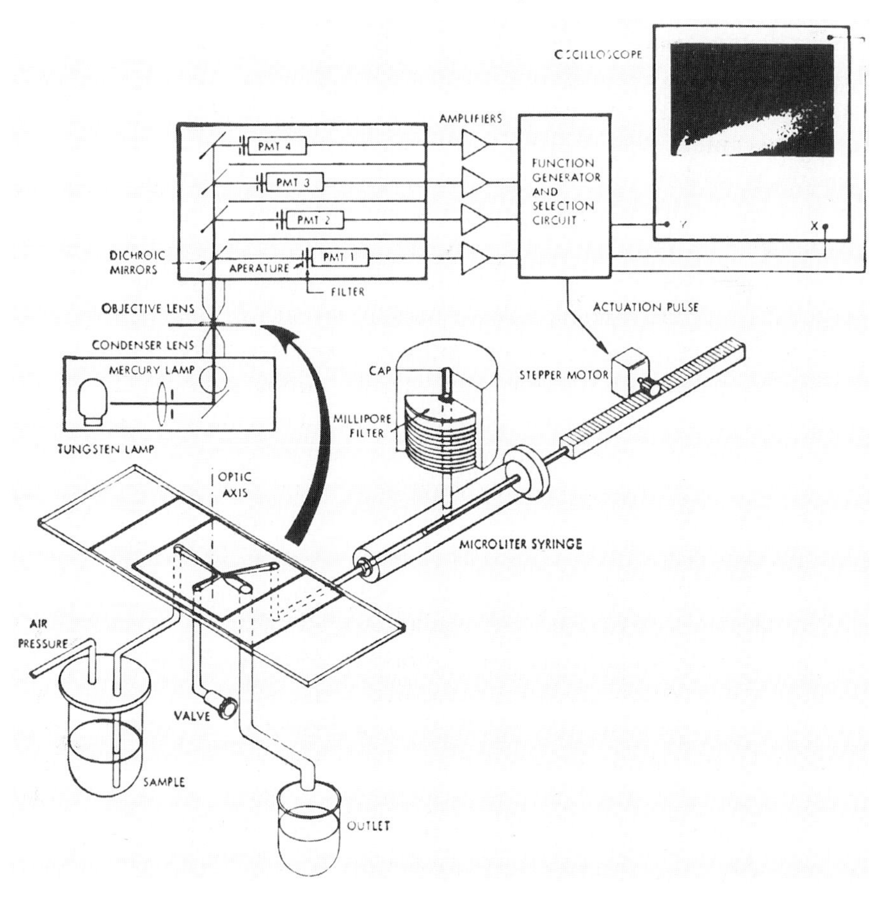


**Fig. 1.2.** A diagram of Kamentsky's original flow sorter. From Kamentsky and Melamed (1967).

electrical resistance ("Coulter volume"). Anyone looking at erythrocytes under the microscope cannot help but be impressed by the remarkable structural uniformity of these cells; Fulwyler wondered if the bimodal Coulter volume distribution represented differences between two classes of these apparently very uniform cells or, alternatively, whether the bimodal profile was simply an artifact based on some quirky aspect of the electronic resistance measurements. The most direct way of testing these two alternatives was to separate erythrocytes according to their electronic resistance signals and then to determine whether the separated classes remained distinct when they were re-analyzed.

The technique that Fulwyler developed for sorting the erythrocytes combined Coulter methodology with the ink jet technology being developed at Stanford University by RG Sweet for running computer printers. Ink jet technology involved the vibration of a nozzle so as to generate a stream that would break up into discrete drops (think of shaking a garden hose) and then the charging and grounding of that stream at appropriate times so as to leave indicated drops, as they broke off, carrying an electrical charge. For purposes of printing, those charged drops of ink could then be deflected to positions on the paper as required by the computer print messages. Fulwyler took the intellectual leap of combining this methodology with the Coulter flow technology; he developed an instrument that would charge drops containing suspended cells, thereby allowing deflection of the cells (within the drops) as dictated by signals based on the cell's measured Coulter volume.

The data from this limited but ingenious experiment led to a conclusion that with hindsight seems obvious: Erythrocytes are indeed uniform; when cells are sorted according to their electrical resistance, the resulting cells from one class or the other still show a bimodal distribution when re-analyzed for electrical resistance profile. The bimodal "volume" signal from erythrocytes was therefore artifactual—resulting in part from the discoid (nonspherical) shape of the cells. The technology developed for this landmark experiment is the essence of all the technology required for flow sorting as we now know it. That experiment also, unwittingly, emphasized two other aspects of flow cytometry that have remained with us to this day: First, there has been a continual (and usually beneficial) tension between scientists who are fascinated by flow technology for its own sake and scientists who are interested in answers to what they see as useful questions. Second, flow cytometrists still need to be continually vigilant because signals from cells (particularly signals that are related to cell volume) are subject to artifactual influences and may not be what they seem. (Fulwyler's 1965 paper actually describes the separation of mouse from human erythrocytes and the separation of a large component from a population of mouse lymphoma cells; the experiments on the bimodal signals from red cells have been relegated to flow history.)

In 1953, Crosland-Taylor, working at the Middlesex Hospital in London, had noted that

> attempts to count small particles suspended in fluid flowing through a tube have not hitherto been very successful. With particles such as red blood cells the experimenter must choose between a wide tube which allows particles to pass two or more abreast across a particular section, or a narrow tube which makes microscopical observation of the contents of the tube difficult due to the different refractive indices of the tube and the suspending fluid. In addition, narrow tubes tend to block easily.

In response to this dilemma, he applied the principles of laminar flow to the design of a flow system. A suspension of red blood cells was injected into the center of a faster flowing stream, thus allowing the cells to be aligned in a narrow central file within the core of the wider stream preparatory to electronic counting. This principle of hydrodynamic focusing was pivotal for the further development of the field.

In 1969, Marvin Van Dilla and other members of the Los Alamos group reported development of the first fluorescence-detection cytometer that utilized the principle of hydrodynamic focusing and, unlike the microscope-based systems, had the axes of flow, illumination, and detection all orthogonal to each other; it also used an argon laser as a light source (Fig. 1.3). Indeed, the configuration of this instrument provided a framework that could support both the illumination and detection electronics of Kamentsky's device as well as the rapid flow and vibrating fluid jet of Fulwyler's sorter. In the initial report, the instrument was used for the detection of fluorescence from the Feulgen-DNA staining of Chinese hamster ovary cells as well as of their Coulter volume; however, the authors "anticipated that extension of this method is possible and of potential value." Indeed, shortly thereafter the Herzenberg group at Stanford published a paper demonstrating the use of a similar cytometer to sort mouse spleen and Chinese hamster ovary cells on the basis of their fluorescence due to accumulation of fluorescein diacetate. These instruments thus led to systems for combining multiparameter fluorescence, scatter, and "Coulter volume" detection with cell sorting as we now know it.

These sorting cytometers began to be used to look at ways of distinguishing and separating white blood cells. By the end of the 1960s, they were able to sort lymphocytes and granulocytes into highly purified states. The remaining history of flow cytometry involves the elaboration of this technology, the exploitation of the cytometers for varied applications, and the collaboration between scientists and industry for the commercial

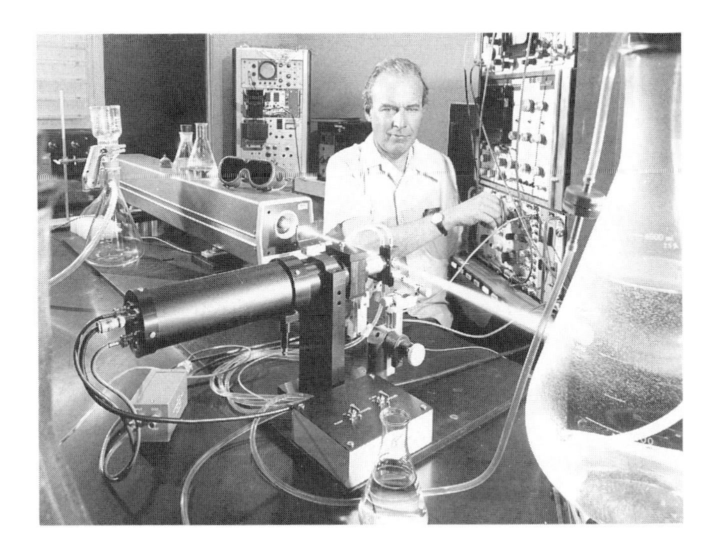

**Fig. 1.3.** Marvin Van Dilla and the Livermore flow sorter in 1973. Photograph courtesy of the Lawrence Livermore National Laboratory.

**Fig. 1.4.** Bernard Shoor (left) and Leonard Herzenberg at Stanford University with one of the original Becton Dickinson flow cytometers as it was packed for shipment to the Smithsonian Museum. Photograph by Edward Souza courtesy of the Stanford News Service.

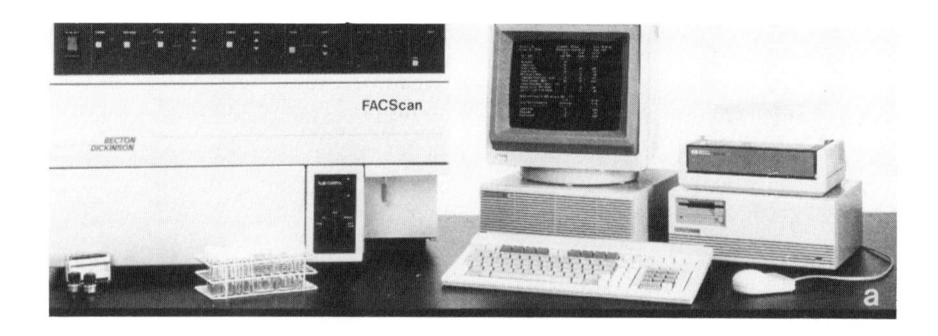


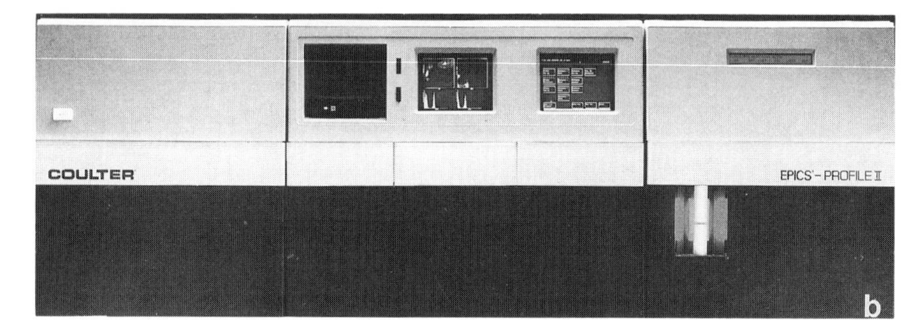


**Fig. 1.5.** Two user-friendly cytometers. **a:** A Becton Dickinson FACScan. **b:** A Coulter Profile.

production of cytometers as user-friendly tools (Fig. 1.4). At the same time that these instruments began to be seen as commercially marketable objects, research and development continued especially at the Los Alamos and Livermore National Laboratories, but also at smaller centers around the world. At these centers, homemade instruments continue to indicate the leading edge of flow technology. At the present time, this technology is moving simultaneously in two directions: on the one hand, increasingly sophisticated instruments are being developed that can measure and analyze more aspects of more varied types of particles more and more sensitively and that can sort particles on the basis of these aspects at faster and faster rates; on the other hand, a different type of sophistication has streamlined instruments (Fig. 1.5) so that they can become user-friendly and essential equipment for every laboratory bench.

## FURTHER READING

Throughout this book, the "Further Reading" references at the end of each chapter are intended to point the way into the specific literature related to the chapter in question. At the end of the book, "General References" will direct readers to globally useful literature. Titles in bold at the end of each chapter are texts that are fully cited in the General References at the end.

Coulter WH (1956). High speed automatic blood cell counter and size analyzer. Proc. Natl. Electronics Conf. 12:1034-1040.

Crosland-Taylor PJ (1953). A device for counting small particles suspended in a fluid through a tube. Nature 171:37-38.

Dittrich W, Göhde W (1969). Impulsfluorometrie dei einzelzellen in suspensionen. Z. Naturforsch. 24b:360-361.

Fulwyler MJ (1965). Electronic separation of biological cells by volume. Science 150:910-911.

Herzenberg LA, Sweet RG, Herzenberg LA (1976). Fluorescence-activated cell sorting. Sci. Am. 234:108-115.

Hulett HR, Bonner WA, Barrett J, Herzenberg LA (1969). Cell sorting: Automated separation of mammalian cells as a function of intracellular fluorescence. Science 166:747-749.

Kamentsky LA, Melamed MR (1967). Spectrophotometric cell sorter. Science 156:1364-1365.

Kamentsky LA, Melamed MR, Derman H (1965). Spectrophotometer: New instrument for ultrarapid cell analysis. Science 150:630-631.

Moldavan A (1934). Photo-electric technique for the counting of microscopical cells. Science 80:188-189.

Van Dilla MA, Trujillo TT, Mullaney PF, Coulter JR (1969). Cell microfluorimetry: A method for rapid fluorescence measurement. Science 163:1213-1214.

Chapter 1 in **Melamed et al.** and Chapter 3 in **Shapiro** are good historical reviews of flow cytometry.

# 2

# Setting the Scene

As mentioned in the previous chapter, flow cytometry has been moving in two directions at once. The earliest flow cytometers to be used in labs either were homemade Rube Goldberg (Heath Robinson in U.K. terminology) instruments or, a few years later, were equally complex, unwieldy commercial machines. These were expensive; and they were unstable and therefore difficult to operate and maintain. For these reasons, the cytometer tended to collect around itself the trappings of what might be called a flow facility. By this I mean a group of scientists, technicians, students, and secretaries as well as a collection of centrifuges, microscopes, and computers that all revolved around the flow cytometer at the hub. If a scientist or clinician wanted the use of a flow cytometer to provide some required information, he or she would come to the flow facility, discuss the experimental requirements, make a booking, and then return with the prepared samples at the allotted time. The samples would then be driven through the cytometer by a dedicated and knowledgeable operator. Finally, depending on the operator's assessment of the skill of the end-user, a number, a computer print-out, or a computer disk would be handed over for analysis.

Such flow facilities still exist. Their existence reflects a need for the power and adaptability of state-of-the-art instrumentation, but also recognition of its high cost, its requirement for skilled maintenance and operation, and the fact that many users from many departments may each require less than full-time use of such an instrument. Such centralized facilities may have more than one cytometer. The trend now is frequently to have one or two sophisticated instruments for specialized procedures accompanied by one or more simpler cytometers as routine work horses.

Such facilities usually have dedicated operators who run the instruments and ensure that they are well-tuned for optimum performance. In addition, there may be a network of computers so that flow data can be analyzed and re-analyzed at leisure—away from the cytometer and possibly in the scientist's own laboratory.

Increasingly, such institutional flow facilities are also becoming general cytometry facilities (not just *flow* cytometry facilities); the use of image analysis microscopes for cell analysis has proven to be a technique that complements flow work. Similar stains are used in both flow and image analysis systems: flow cytometry is based on analysis of light scatter and staining from a continuous sample of cells as they pass a detector; image analysis studies the distribution of light signals emanating from a single stationary cell as it changes with time under the scrutiny of the detector. A cytometry facility may also provide general cell equipment such as centrifuges and fluorescence microscopes to aid in cell preparation.

The funding of such central facilities often involves a combination of institutional support, research grant support, and charges to users. The charges to users may vary from nominal to exorbitant and may need to support everything from consumables like computer paper and test tubes to the major costs of laser replacement, service contracts, and staff salaries. It should be clear that, with the increasing complexity and variety of instrumentation and with the need for data organization, booking systems, and financial accounting, the running of these flow facilities can become an administrative task in itself. Therefore a "corporate identity" often evolves, with logos designed, meetings organized, newsletters written, and training sessions provided.

Just at the moment when such elaborate flow facilities were beginning to see themselves as the source of all knowledge, the technology of flow cytometry began to move in a new direction. Slowly users and manufacturers both began to realize that different laboratories have different instrumentation needs. A sophisticated sorter might be necessary for certain applications, but its demands for daily alignment and skilled maintenance are time-consuming and its research capabilities might be superfluous for routine processing of, for example, clinical samples. Instead of simply continuing to become larger, more complicated, and more powerful, some flow cytometers started to become available that were small, accessible, and friendly, albeit less powerful.

It is actually the commercial marketing of user-friendly instruments, designed to a great extent with the clinical laboratory in mind, that has changed all sorts of things that we were beginning to regard as intrinsic features of flow research. These new instruments are less expensive and, being somewhat limited in scope, can be compact, easily installed, rela tively stable, and maintained with little human intervention. Therefore users who previously found that they had been walking miles each week in daily treks to their institution's flow facility suddenly realized that it might be possible to consider having their own instrument (it would fit in nicely right next to the balance and the pH meter on the lab bench). This is, in fact, just what has been happening. With great success, these friendly cytometers are being rapidly incorporated into ordinary research laboratories as well as into routine hospital service laboratories. The success of the "black box" technology can most be judged, indeed, by the fact that these instruments, although still relatively expensive, have started to be taken for granted.

My prejudice on this issue should, of course, be obvious. If I believed that flow data could be acquired and analyzed appropriately by people with no awareness of the limitations and assumptions inherent in the technique, I would not have written this book. The actual operation of these small cytometers has been vastly simplified compared with that of the original research instruments; it certainly can be said that the new wave of cytometers has the potential for making flow analysis a great deal more accessible. A serious concern, however, is that the superficial simplicity of the instruments may lull users into a false sense of security about the ease of interpretation of the results; the basis for this concern is particularly clear in the medical community, where clinicians have been conditioned to expect that laboratory reports will contain unambiguous numbers and may not be accustomed to the need for an intellectual framework in which to interpret those numbers.

Moreover, a funny thing has been happening. Those very scientists who were pleased to be able to buy and run their own cytometers in order to be released from the tyranny of the centralized flow facility are now, from the skills acquired by maintaining their own cytometer, beginning to develop a desire for wider horizons. Not realizing, initially, what the sophisticated instruments could do beyond routine analytical tasks, they now are heading, at least occasionally, back to core flow facilities for the adaptability, sorting, and multiparameter analyses that may only be possible with state-of-the-art instrumentation. So what originally appeared to be a rift in the field of flow cytometry that would eventually isolate the "high tech" from the "routine analysis" contingents has turned out to be more circular and less rigid than one might have predicted. Routine cytometers have begun to be welcomed into flow facilities and used there to alleviate the work load on their sophisticated cousins; and the presence of routine cytometers in ordinary research and clinical labs has, in many cases, whetted the desire for a higher order of analysis that requires state-of-the-art technology.

# 3

# Instrumentation: Beyond the Black Box

I want first to clear up some confusion that results simply from words. A flow cytometer, despite its name, does not necessarily deal with cells; it deals with cells quite often, but can also deal with chromosomes or latex beads or with any other particles that can be suspended in a fluid. Although flow cytometers were originally developed to sort particles, many instruments proudly calling themselves *flow cytometers* are in fact not capable of sorting. The acronym *FACS* is even applied to some of these nonsorting machines, misleading though this is. Flow cytometry might be broadly defined as a system for measuring and then analyzing the signals that result as particles flow in a liquid stream through a beam of light. Flow cytometry is, however, a changing technology, and defining it is something like capturing a greased pig; the more tightly the pig is grasped, the more likely it is to wriggle free. In this chapter, I describe the components that make up a flow cytometric system in such a way that we will not need a definition, but will know one when we see it.

The common elements in all flow cytometers are a light source; an optical bench to focus and direct that light; fluid lines and controls to direct and regulate the flow of a liquid stream containing particles through the focused light beam; an electronic network for measuring the intensity of light signals and recording them (and also, if the instrument has sorting capability, for initiating sorting decisions to charge and deflect particles); and a computer system for analyzing these light signals by correlating them with each other. In this chapter I describe the optical bench, fluid systems, electronic signal processing, and sorting controls; in the next, I will discuss computing and data analysis strategies; and in Chapter 5, light sources will be described.

## THE OPTICAL BENCH

An optical bench is simply a table that does not wobble. A flow cytometer's optical bench may be visible at the back or may be incorporated behind a closed door; in either case, it provides a stable surface that fixes the light source in rigid alignment with the objects being illuminated. What this means is that, if the bench is moved, the light source and the object of illumination will move in synchrony so that alignment between the two does not change. The reason that the users of a flow cytometer should know about optical benches is, simply, that this knowledge serves to remind them that signals from a particle can vary beyond recognition if this alignment varies even slightly.

Figure 3.1 is a diagram of the components that sit on the optical bench of a flow cytometer. If we follow the light path from the beginning, we can see that the light, after it leaves the laser source, is focused through a lens into a beam of about 50 μm cross-sectional diameter as it approaches the

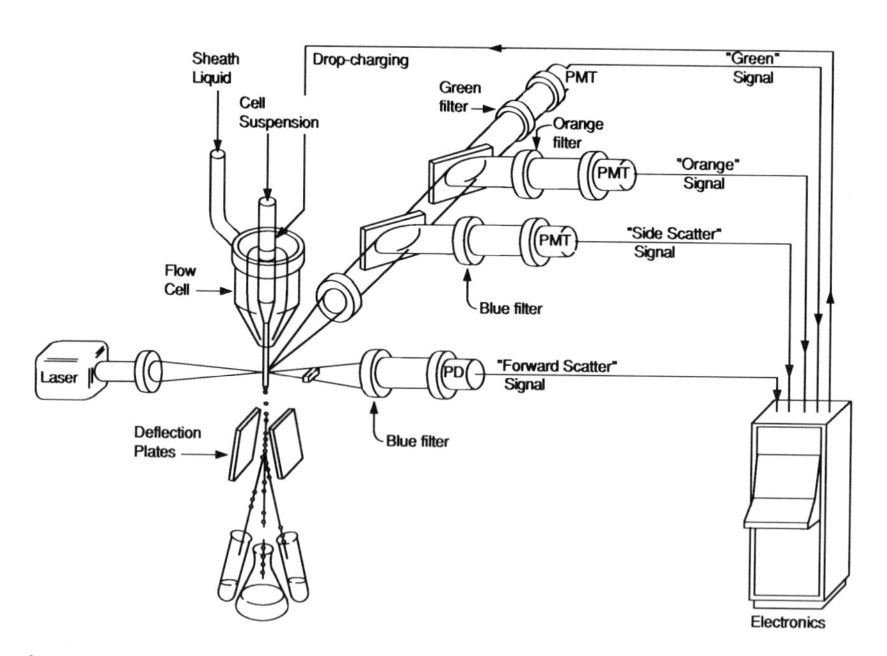


**Fig. 3.1.** Components on the optical bench of a generalized "four-parameter" flow cytometer. Figure adapted from Becton Dickinson Immunocytometry Systems.

liquid stream (lenses that produce ellipsoidal rather than circular beams are used for some applications). A stream of 50–150 μm diameter flows perpendicularly to the beam. The lens responsible for focusing the light beam must ensure that the stream is intersected by that light beam in such a way that the core of the stream is uniformly illuminated by the light. The point at which stream and light beam intersect (Fig. 3.2) is called the *analysis point*, *observation point*, or *interrogation point*. If the light beam and the stream are not perfectly and squarely aligned with each other, then cells within the stream will be erratically illuminated and will give off erratic light signals. Although the reasons for imperfect alignment may have to do with poor adjustment of the focusing lens or light source, the stability of the optical bench is, on the whole, reliable. On a day to day basis, poor alignment is more likely to result from shifts in the fluid stream.

Surrounding the observation point are lenses that collect light as it emerges after its intersection with the stream. This emerging light constitutes the *signal*. It is focused onto photodiodes or photomultiplier tubes (the latter are more sensitive to dim light than the former, but both can be called *photodetectors* to avoid confusion) that convert the light signal into an electrical impulse. The intensity of the electrical impulse is proportional to

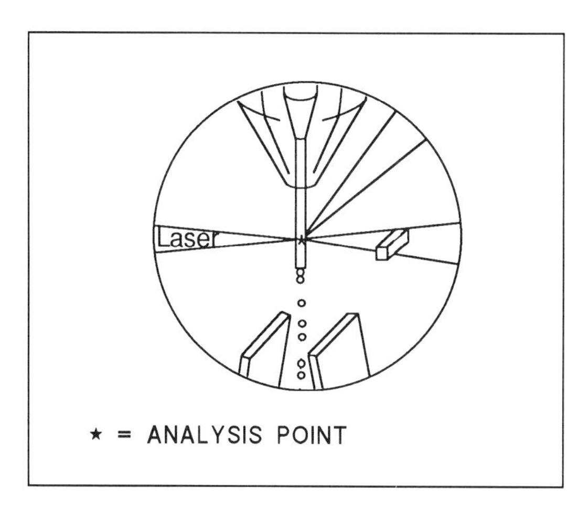


**Fig. 3.2.** The analysis point. Alignment between the illuminating beam and fluid stream are critical in determining the characteristics of the resulting forward and right-angle signals.

the intensity of the light impinging on the detector. Each photodetector has, in front of it, a light filter of a specific color; the job of the filter is to ensure that each photodetector only "sees" light of the color transmitted through its own filter. If a green filter is in front of photodetector number one (pd1) and an orange filter is in front of photodetector number two (pd2), then pd1 will respond with an electrical impulse if green light emerges after intersecting the stream, pd2 will respond with an electrical impulse if orange light emerges, both detectors will respond if white light emerges, and neither will respond if blue, or yellow, or no light appears.

The standard configuration for photodetectors around the observation point in a flow cytometer is three, four, or five photomultiplier tubes at right angles to the illuminating beam. Because they are at the side and not in direct line of the illuminating beam, no light will hit these detectors unless something in the stream causes the illuminating beam to be deflected to the side or unless something in the stream is in itself a source of light in that direction. One of these detectors is usually fitted with a filter of the same color as the illuminating light (blue is the usual color here); it will register any illuminating (blue) light that is bounced to the side from the surface of a rough particle in the stream. The rougher or more irregular or granular a particle is, the more it will scatter the illuminating beam to the side. The intensity of this so-called side scatter (SSC) light (defined as light of the same color as the illuminating beam that is scattered by a particle to an angle of 90° from the illuminating beam) is therefore related to the texture of the particles flowing in the illuminating light path; it is sometimes referred to as a *granularity signal* or an *orthogonal light scatter signal.* Granulocytic blood cells with their irregular nuclei, for example, have a much more intense SSC than do the more regular lymphocytes or erythrocytes.

The several other photomultiplier tubes that sit at right angles to the illuminating beam are usually fitted with green, orange, and red filters—depending on how many of these detectors are present in a particular instrument. If the illuminating beam is blue, then there is no way that red, orange, or green light will emerge from the analysis point unless a particle in the stream is itself generating that red, orange, or green light. The usual way that a particle emits light is either because it contains endogenous fluorescent compounds or because a scientist has stained it with a fluorescent stain. A cell stained with a fluorescent stain will, when illuminated from one direction by a beam of blue light, emit in all directions light of a different color (the color depending on the fluorescent stain used). It is this fluorescence that will be registered on one or another of the right angle photomultiplier tubes—depending on the color of the fluorescence and the filter in front of the tube. In this way, the signal from each photomultiplier tube indicates the presence of a particular stain on a particle.

The photodiode that sits in direct firing line of the illuminating beam gives out signals that are responsible for much misunderstanding. Rather than simply detecting the amount of the blue illuminating beam that gets through after bombarding a particle, this forward angle photodetector is fitted with a so-called obscuration bar in front of it. The function of this narrow metal strip is to block the illuminating beam so that it cannot reach the detector. It may be wondered why we bother to place a photodetector in this position on the optical bench if we are then going to prevent any light from hitting it by blocking that light with a strip of metal. The interesting point about this technique is that any light that has been bent as it passes through a particle will manage to escape around this bar and generate a signal. Depending on the width of the bar, the angle of bending required to generate a signal is usually about 0.5°. This so-called forward scatter (FSC) or forward angle light scatter (FALS) signal (defined as light of the same color as the illuminating beam that is bent to a small angle from the direction of that original beam) is sometimes called a *volume* signal. It is undoubtedly related to volume, but it is also related to other factors such as the refractive index of the particles. A particle with a refractive index very different from that of the surrounding medium will bend more light around the bar than will a particle with a refractive index closer to that of the stream. It is true that a large particle will bend more light than a small particle of the same refractive index; however, confusion inevitably results when the term *volume* is used carelessly to describe this forward angle scatter.

For example, anyone used to looking at cells under a microscope will say that dead cells appear larger than living cells of the same type. It is therefore something of a surprise to find that dead cells appear "smaller" than living cells in a flow cytometer. Of course, they are not actually smaller, they simply give a dimmer forward scatter signal than living cells because they have leaky outer membranes and the refractive index of their contents has for this reason become more like the refractive index of the surrounding stream. They therefore bend less light into the FSC detector than a viable cell does. To give one more example (just to hammer the message home), erythrocytes, despite their uniform volume, give a broad range of forward scatter signals in a flow cytometer because their effective "volume" varies depending on their orientation in the stream. But anyone who remembers

Fulwyler's experiments with red cells in a Coulter counter should not find this fact too surprising.

Since I mentioned Coulter volume measurements, I should add, for the sake of completeness, that some flow cytometers (most notably the Becton Dickinson FACS Analyzer) have a volume signal that is based on the Coulter-type measurement of electrical resistance. This is not in any way related to the FSC signal discussed in the paragraphs above. The Coulter-type volume signal is proportional to the volume of a particle (as well as to its electrical resistance characteristics). The FSC signal is proportional to the cross-sectional area of a particle (as well as to its refractive index). Some state-of-the-art cytometers actually record both kinds of signals from particles; the ratio between the two varies for different types of particles, and this can be instructive. The take-home-message is, therefore, that both of these kinds of signals give information about the physical characteristics of a particle, but neither tells much that we can count on about the true volume of that particle. The news, however, is not all bad. Although we may not know the exact physical meaning of a forward scatter signal, this scatter characteristic can allow us to distinguish different classes of particles from each other, and this can be useful.

By way of recapitulation, before we move on to electronics, recall simply that, when a focused light beam hits a particle in a stream of fluid, the light is affected in several ways. It can be reflected and/or refracted; it can also be converted to a different color if it has been absorbed by a fluorescent chemical. The light that emerges after hitting a particle may register on one or more of the available photodetectors. The photodetectors each measure some specific aspect of the emerging light because of their positions, the color of their filters, or the presence of an obscuration bar. In a typical configuration, two of the photodetectors measure forward angle scatter light and side scatter light in order to provide some information about the physical characteristics of the particle (sometimes called *volume* and *granularity*); and two or more photodetectors are equipped with colored filters to provide information about the fluorescent light being emitted by the particles. These four or more characteristics registered on the photodetectors as each particle passes through the light beam are known as the *measured parameters* in the flow cytometry system. An instrument may, for example, be called a *three-parameter cytometer* or an *eight-parameter instrument* depending on how many photodetectors are arranged around the analysis point. Most commercial instruments now have a five or six parameter configuration.

## THE FLUID SYSTEM

The fluid system (or fluidics) in a flow cytometer is likely to be ignored until it goes wrong. If it goes wrong disastrously, it can make a terrible mess. If it goes wrong with subtlety, it may turn a good experiment into artifactual nonsense without anyone ever noticing. On the assumption that the more disastrous problems can be solved by a combination of plumbing and mopping (both essential skills for flow cytometrists), I will concentrate on the more subtle aspects of fluid control. Nevertheless, the potential hazard of working at the same time with volumes of water and with a high-voltage source should never be far from the mind of anyone working with a water-cooled laser or with a sorting cytometer with high-voltage stream deflection plates.

The flow on a flow cytometer begins (Fig. 3.3) at a reservoir of liquid, called the *sheath fluid*. Sheath fluid will provide the supporting vehicle for directing particles through the flow system. The sheath fluid reservoir is pressurized, usually with pumped room air, to drive the sheath fluid through a filter to remove extraneous particles and then through plastic tubing to the flow cell. The sheath fluid is usually buffer of a composition that is appropriate to the types of particles being analyzed. For leukocytes or other mammalian cells, this usually means some sort of phosphate-buffered saline solution. Other particles have other preferences. However, after reading the rest of this section, you will realize that the particles flowing through the cytometer are not ever actually in contact with the sheath fluid from this reservoir until after they have been analyzed. Therefore, for nonsorting applications, it does not generally matter what is in the sheath reservoir. At least one company producing cytometers recommends that distilled water be used in the sheath tank. Distilled water certainly is cheap and easy. Tap water is even cheaper and easier. My main reservation about recommending water instead of buffer for the sheath supply is that the sheath supply, in some instruments, may sometimes drip back into the sample before it has been analyzed. If this is the case, then it would obviously be better to have the sheath fluid compatible with the well-being of the cells in question. (There may also be some less common cytometer applications in which rapid diffusion of small molecules between cells and sheath might critically affect results; if stains are in rapid equilibrium between cellular components and medium, then you may want stain in the sheath fluid in order to maintain appropriate concentrations in the cells.) Sorting is different. During sorting, the particles end up suspended in the

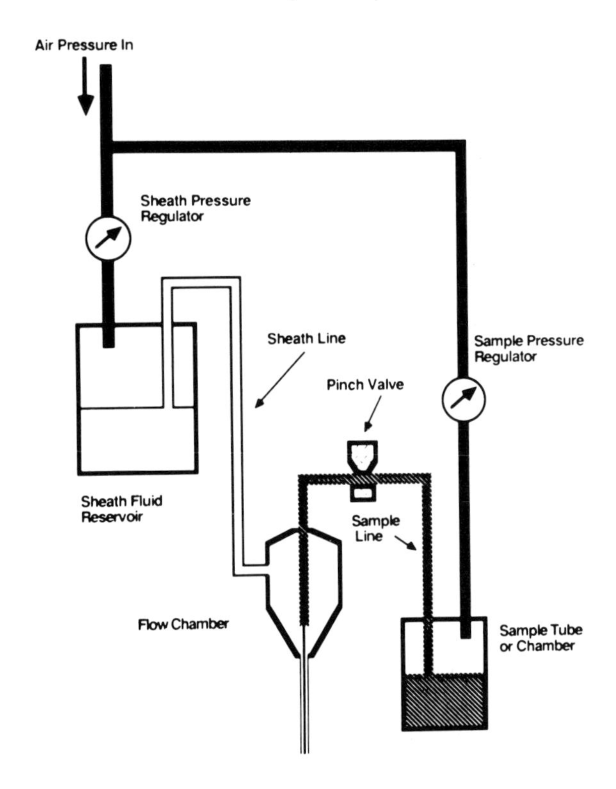


**Fig. 3.3.** The fluidics system. From Carter and Meyer (1990).

sheath fluid for further culture or analysis; in that case, the composition of the sheath fluid is obviously important if the particles are at all sensitive in their requirements. And the other thing about sorting is that the sheath fluid must have a high enough electrolyte concentration so that it is able to carry a charge; distilled water will not work here.

Different instruments employ different strategies for getting the sample with suspended particles into the system. Some instruments use small test tubes that form a tight seal around an O-ring on a manifold. The manifold delivers air to the test tube, thus pushing the sample up out of the test tube and through a plastic line to the flow cell. Other instruments use a motor-driven syringe to remove a volume of sample and then inject it slowly into the flow cell. Depending on the instrument, there may be a greater or lesser degree of operator control over the rate of flow. The amount of pressure driving the sample through the system may affect the uniformity of alignment between the particles and the illuminating light beam as the particles move through the flow cell. Low pressure is less likely to cause perturbation of the stream profile and of the position of the particles within that stream. Empirically, if increasing the pressure pushing the sample through the flow cell causes undue broadening or wavering of signals, the pressure is probably excessive.

If particles flow too slowly through the cytometer, people start to make bad jokes about how microscopes cost less and are quicker. Since increasing the pressure may not be possible and, even if it is, is probably only a good idea within the reasonable limits discussed above, the best way of getting particles to flow at reasonably fast rates is simply to make up the original sample with particles at a reasonably high concentration. A million cells per milliliter is often about right; $10^5$ cells per milliliter is beginning to be low enough to test one's patience; $10^4$ cells per milliliter is probably too low a concentration to be worth analyzing.

If you have few cells, make them up in a small volume (you will know how small a volume your system can handle). If the cells end up being too concentrated, they may flow too fast—but you can always dilute them on the spot and run the sample again. You may wonder why too rapid a flow is a source of problems. Faster seems as if it should always be better (especially around 5:00 PM). However, if particles flow too rapidly through the flow cell, there may be difficulty separating their signals: Two cells may arrive in the illuminating beam at the same time and may be measured together as if they were a single particle, and thus errors will result (again, look for broad and wavering signals). At a rate slow enough not to cause these coincidence events, the computer may still not be able to count fast enough to record all the information. This may be your problem if you find that you run out of sample before you have been able to analyze the number of particles you think you put into the sample tube. Most cytometers seem to be quite happy to look at particles that are flowing at a rate of about 1,000 particles per second. At least some computing systems for data storage seem to miss particles if they flow at a rate much faster than this. Sorting, for other reasons, has its own restrictions on particle flow rate.

Aside from concentration, another problem with samples is that the particles may be the wrong size. If they are too small, they may not be distinguishable from noise; nevertheless, bacteria and picoplankton and other bits and pieces of about 1 μm size or smaller are probably analyzable in at least some well-tuned cytometers. But, if the particles are too large, they will clog the tubing or the flow cell. If the fluid system is fully clogged, it may be difficult to get things flowing again; if, however, it is only partially clogged, cells may flow but that critical alignment between stream and light beam may be skewed, thus causing artifactual signals. Most experienced flow cytometrists recommend filtering any samples that are likely to contain large or clumped material before attempting to run them through the instrument. Nylon mesh of specified pore size works well (35 μm mesh is appropriate for most applications).

The exact size of particle that will be large enough to cause clogging problems depends on the diameter of the orifice of the nozzle or flow cell being used. This brings us to the next stop downstream in our following of the flow in this flow cytometer. The term *nozzle* may be used interchangably with *flow cell* or *flow chamber*, although the terms derive from different engineering designs for the best way of delivering particles into the sheath fluid and thence to the analysis point where they are illuminated by the light. In some systems, the stream emerges from a nozzle with a small orifice where it is then intersected by the light beam in the open air. In other systems, the stream is directed (either upward or downward) through a narrow, optically clear, flow cell or cuvette; the particles are hit by the light beam while they are still within this flow chamber. In still other systems, a nozzle forces the stream at an angle across a glass coverslip (Fig. 3.4). All systems have their advocates—the positive and negative considerations are based mostly on ideas about signal noise, stream turbulence, and the control of drop formation for sorting. There is no clear favorite, and purchasers of commercial instruments usually base their choice of cytometer on factors other than flow cell design and then live with the design they get.

The main point of concern on a daily basis is the avoidance of blockages. Most nonsorting cytometers are tolerant of large material (perhaps 150 μm in diameter), but sorting instruments are more restricted in nozzle size for reasons that are discussed below in the section on sorting. Although instruments can be modified to sort large cells, most commercial sorting cytometers operate with a stream diameter of between 50 and 100 μm. Particles larger than the nozzle diameter will obviously clog the nozzle. However there is also an additional consideration. Because the light beam is focused to a spot on the stream that will have a cross-sectional diameter considerably smaller than the stream width (and the beam's region of uniform intensity will be considerably less than its total diameter), the

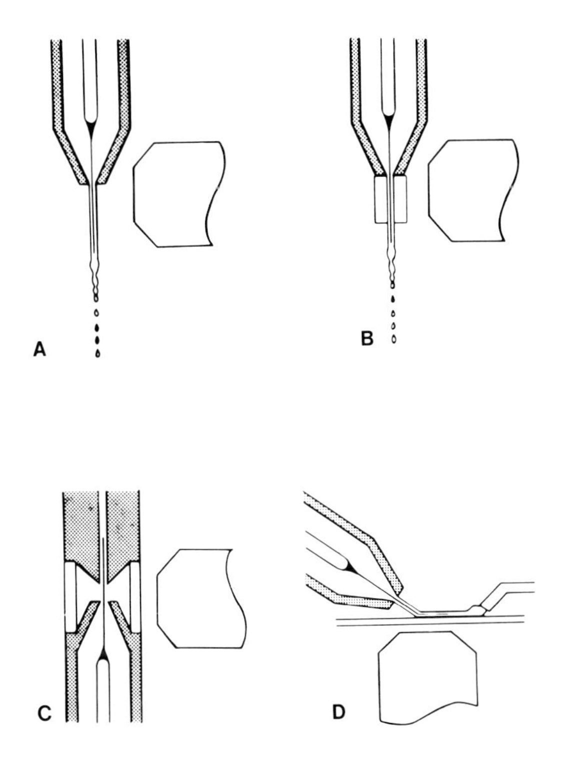


**Fig. 3.4.** Various types of flow chambers. **A** and **B** are designs used in sorting cytometers (in A the analysis point is in air after the stream has left the flow cell; in B analysis occurs within an optically clear region of the chamber itself). **C** and **D** are two designs for nonsorting cytometers (in C the stream flows upward through an optically clear region of the chamber; in D the stream is directed at an angle across a glass coverslip). Adapted from Pinkel and Stovel (1985).

particle's size needs to be considerably less than the nozzle (stream) diameter if it is to be accurately illuminated.

What we now require is a method for keeping the particles in the center of the stream so that they will be uniformly illuminated as they pass through the focused light beam. This is the function of the sheath fluid. The flow cell is the place in the cytometer where the particles from the sample join the fluid from the sheath reservoir. The way in which the sample from the sample line is injected into the sheath stream as they both enter the flow cell is the subject of equations and approximations.

References at the end of this chapter will direct interested readers to mathematical discussion of the hydrodynamics of this subject. For our purposes, it will perhaps be sufficient to note that, as a result of hydrodynamic considerations pertaining to laminar flow, the sample, after injection into the center of the sheath stream, will remain in a central core as it flows within the sheath out through the flow cell. The technical term for this is *hydrodynamic focusing*; flow of a sample stream within the center of a sheath stream is called *coaxial flow*. The exact diameter of that central core is related to, among other things, the rate at which the sample is injected into the sheath stream; a 100 μm stream may, depending on sample injection velocity, have a core width of about 20 μm (see Table 3.1). Because the laminar flow of the stream tends to confine the sample to this central core, there is little mixing of sample with sheath fluid (diffusion of small molecules will occur). The reason that this type of coaxial sample injection suits flow cytometry is that a narrow nozzle that would block easily can be avoided; the particles to be analyzed are nevertheless maintained in alignment in the core so that they progress in single file down the center of the stream. This ensures uniform illumination as long as the central core is narrow and the illuminating beam is focused at the center of the sheath stream. It also ensures that, for the most part, one particle is analyzed at a time. We are now in a position to understand why radical changes in the rate at which the sample is driven through the flow cell will cause changes in the resulting light signals. Changes in sample injection rate cause changes in the diameter of the core; when the size of the core increases, particles are no longer so tightly restricted in their position as they flow past the light beam and illumination may become less uniform (Fig. 3.5). The size of the core

**TABLE 3.1. The Effect of Flow Rate on the Diameter of the Sample Core**

| Flow rate (cells/sec) | Core diameter (μm) | |
|---|---|---|
| | Sample conc: $10^6$ cells/ml | Sample conc: $10^5$ cells/ml |
| 10 | 1 | 3 |
| 100 | 3 | 10 |
| $10^3$ | 10 | 30 |
| $10^4$ | 30 | 100 |

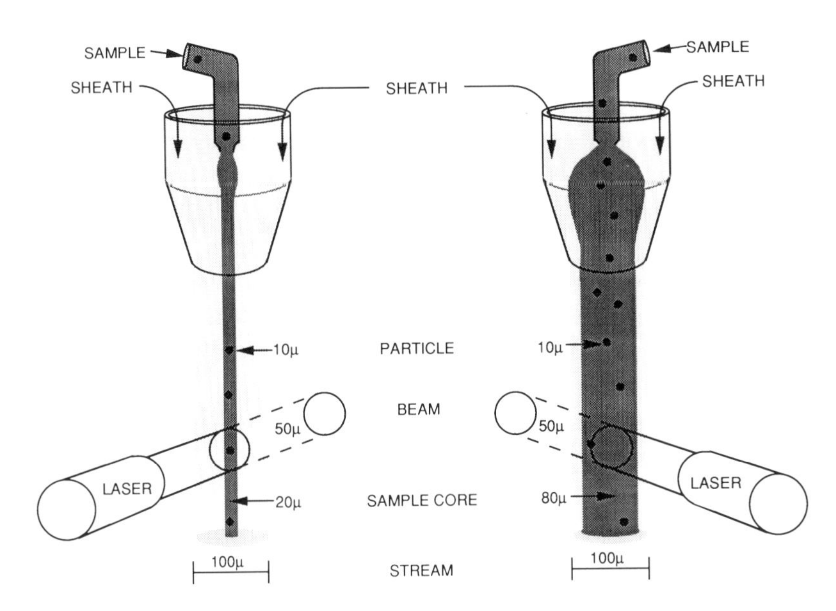


**Fig. 3.5.** The flow of particles within the core of the sheath fluid. When the sample is injected slowly (left), the core is narrow and the particles are confined to the center of the laser beam. When the sample is injected too rapidly (right), the core is wide (somewhat exaggerated in this drawing), and the particles may be illuminated erratically.

often has critical impact in this way on DNA applications where precise analysis is important and non-uniform illumination causes non-uniform fluorescence. The stream diameter may be 100 µm, but if the illuminating beam has a diameter of 50 µm, then the core diameter and the particles within it need to be considerably less than 50 µm to ensure uniform illumination.

In our voyage downstream in the cytometer, we have now almost reached the sea. The sample containing particles has joined the sheath fluid, and within the core of that sheath stream, the particles have, with perfect orthogonal alignment, reached the intersection with the illuminating light beam. At this point, light signals from scatter or fluorescence will be emitted and registered on the photodetectors and, unless sorting is part of the itinerary, the role of the particle is finished and it progresses into the waste container. Once the light has hit the photodetectors, the light signals are converted into electrical signals.

## ELECTRONICS

Even mention of the word *electronics* is enough to strike fear into the hearts of many otherwise perfectly competent people. The circuit boards of a flow cytometer are things of beautiful complexity with which we are not going to concern ourselves. Integrated circuits (chips) do occasionally wear out; if they do, the symptoms can be most peculiar and difficult, even for a trained cytometer engineer, to diagnose (in our lab, we once had side scatter signals masquerading as green fluorescence). The message here is that you should get to know the engineer who services your instrument, and you should be nice to her at every possible opportunity; you will almost certainly need her expertise some day. In this section, however, I will describe just enough of the electronics so that you can understand the control that can be exerted by the operator over the way the light signal given off by a particle reaches the computational circuitry for data analysis.

Photodetectors, as I have said, convert light signals into electrical impulses. The intensity of those electrical impulses are, within the limits of normal operation, related to the intensity of the light signals. But the way in which the electrical impulses are then treated so that they are strong enough to be processed is open to user choice. Photomultiplier tubes have voltages applied to them so that the cascade of electrons resulting from the original light impulse is converted into a sufficiently large current to be measured. Changing the voltage applied to a photomultiplier is one way we have of increasing or decreasing the sensitivity of that tube (if we go too high or too low on the voltage, the response of the tube will no longer be proportional to the amount of light received [the linearity of response can be checked using calibrated fluorescent beads]). The second method we have at our disposal to increase or decrease the sensitivity of our light detection is to change the amplification of that electrical current after it leaves the photodetector. Amplifiers can be either logarithmic or linear and can operate at varying gains. In general, we should be aware of the fact that a logarithmic amplifier allows us to look at light signals over a wide range of intensity; a linear amplifier restricts our sensitive measurements to signals all in the same range.

Thus our two types of controls on the photodetectors are, one, the control of the voltage applied to the detector and, two, our control over the amplifier; our control of the amplifier can take the form of choice of log or linear amplification and also of choice of the precise gain applied. In addition, logarithmic amplifiers often allow an "offset" control, allowing the selection of the intensity range to be analyzed without changing the amplification gain. A comparison with log and linear graph paper is often helpful here. Figure 3.6 uses the graph paper analogy to indicate the effect these voltage and gain choices have on six theoretical signals, with intensities in the relationship of 1:2:10:20:100:200 to each other. By looking at the "graph paper" scale on the lower horizontal axes, it can be seen that, with a linear amplifier, the origin is equal to zero; changing the voltage on the detector or changing the amplifier gain simply stretches or contracts the scale. With a logarithmic amplifier, as with log graph paper, the origin is never equal to zero; changing the offset or the voltage changes the value assigned to the origin; changing the amplifier gain stretches or contracts the scale. The gain setting used with a log amplifier is often referred to as *3 log decades full scale* or *5 log decades full scale*. What this means is that the top of the scale represents an intensity $10^3$ or $10^5$ times as bright as the bottom of the scale (think of log graph paper with 3 or 5 cycles). Some cytometers use log amplifiers that are fixed at a scale encompassing 4 log decades. With other cytometers, the gain on the log amplifier can be varied by the operator to give a range of perhaps 2 to 5 log decades for the full scale.

With a log amplifier, as with log graph paper, signals that are double in intensity relative to each other will be the same distance apart no matter where they are on the scale; in addition, distributions of signals with the same coefficient of variation (CV) will look the same no matter where they appear on a log scale. Neither of these things are true with a linear amplifier or with linear graph paper. Log amplifiers are usually used to analyze fluorescence signals from cells with stained surface markers, because these cells often exhibit a great range of fluorescence intensities. Linear amplifiers are usually used for analyzing the DNA content of cells, because the DNA content of cells does not normally vary by more than a factor of 2 (e.g., during cell division). Linear amplifiers are usually used to analyze forward and side scatter signals, but practice here is apt to vary from lab to lab.

Once the electrical signal from the photodetectors has been amplified, its intensity is then analyzed and this value will be recorded by means of an analog to digital converter (ADC). The role of an ADC is to look at a continuous distribution of signals and divide them up into discrete ranges—much as you would need to do if you wanted to plot the heights of a group of people on a bar graph (histogram). For a height distribution histogram, the first bar might represent the number of people with heights between 4′9.5″ and 4′10.5″ (4′10″ ± 0.5″); the second bar, the number of people with

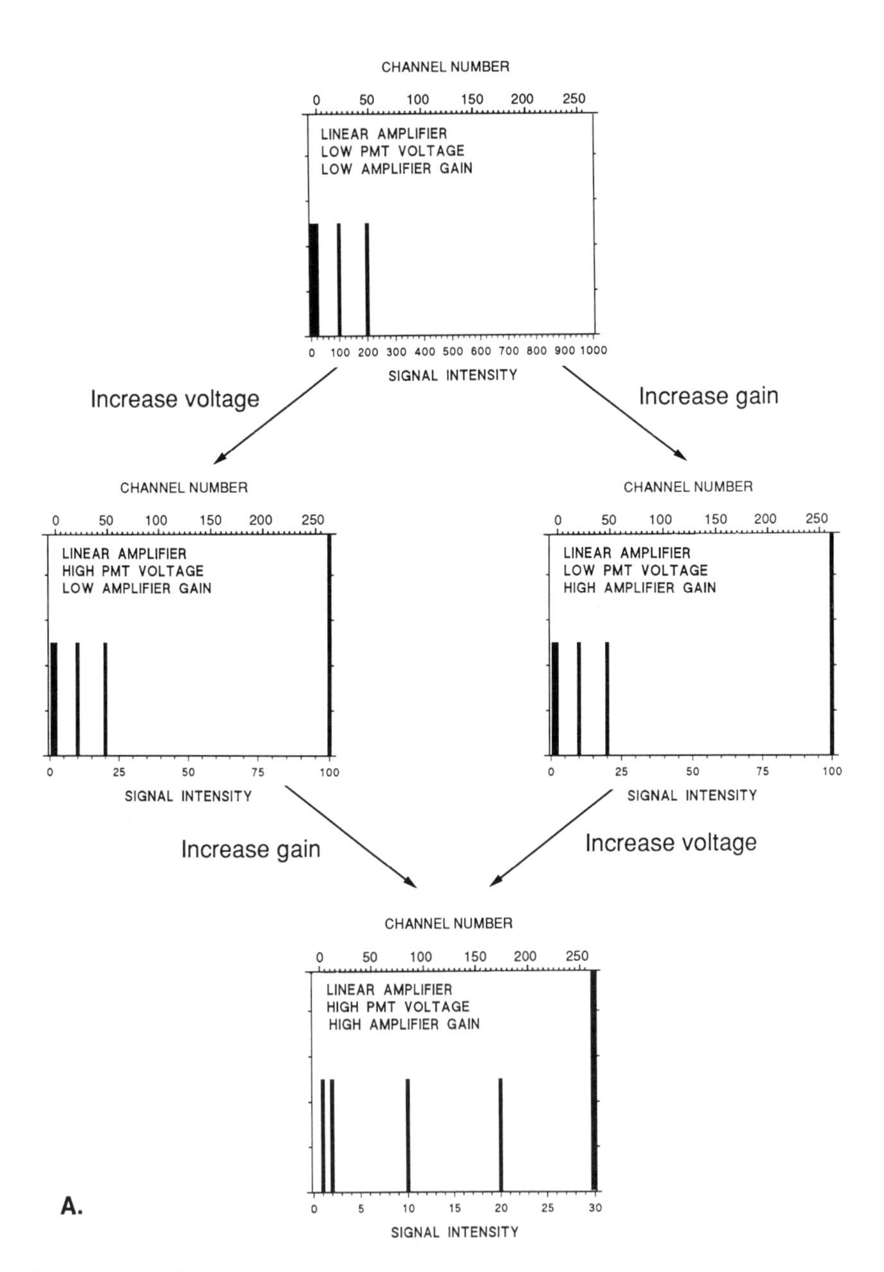


**Fig. 3.6.** The effect of changes in the PMT voltage and amplifier gain on the appearance of six signals with intensities in the relationship of 1:2:10:20:100:200 to each other. **A**: Linear amplification. **B**: Logarithmic amplification.

# LOG AMPLIFIER

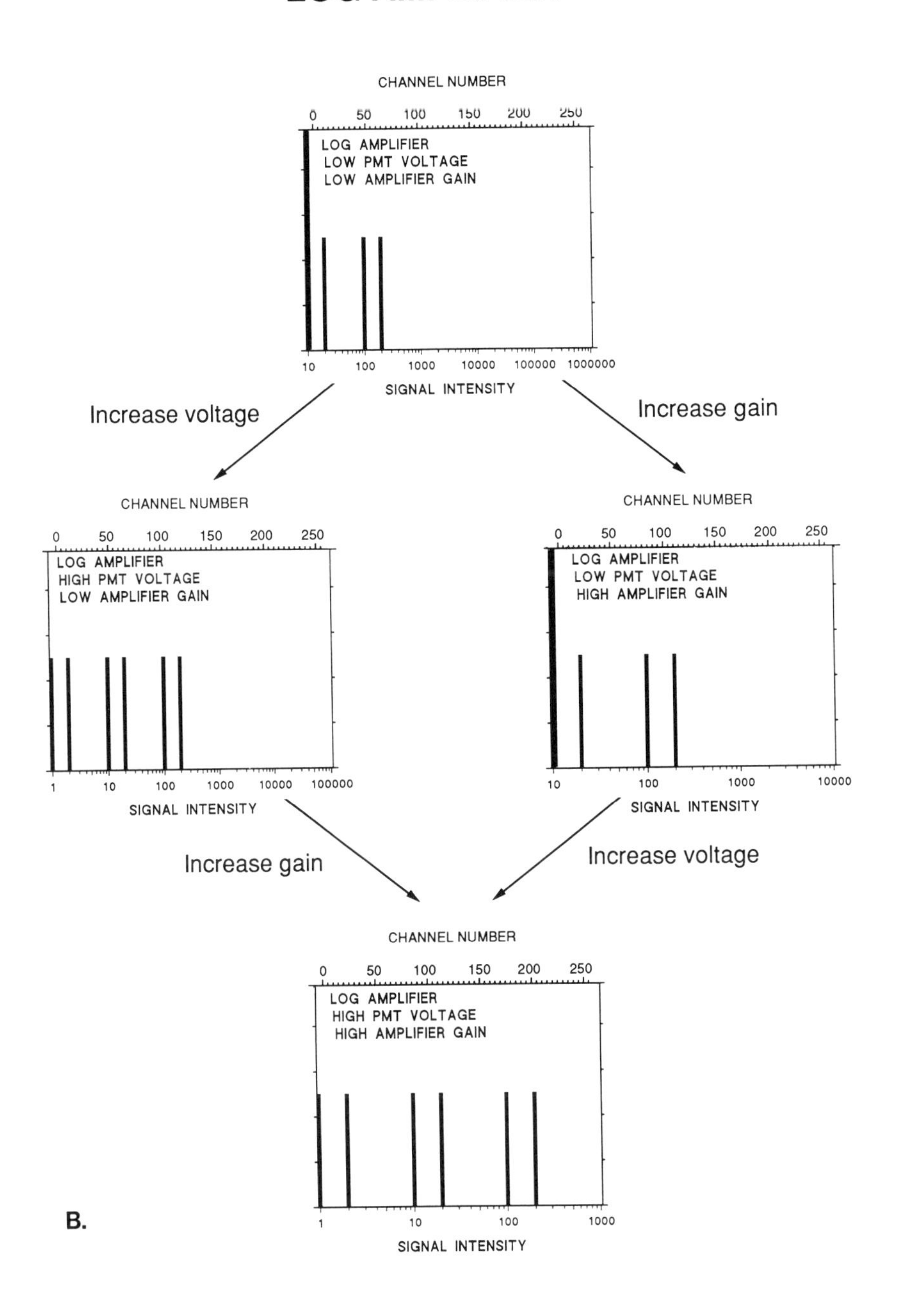

heights between 4′10.5″ and 4′11.5″; and so on (Fig. 3.7). Similarly, the ADC of a flow cytometer divides the electrical signals that it receives from each photodetector into discrete ranges.

Here we run into that term *channel* that flow cytometrists are so fond of. The ADC is divided up into a discrete number of channels; the number of channels is usually either 256 or 1,024. Each channel represents a certain specific light intensity range (like the 17 channels each with a 1 inch range on our height example), and the signal from a cell is recorded in one or another channel depending on the intensity of that signal. It is the combination of photodetector voltage, amplifier gain, and offset that allows the user to set the intensity range represented by each of those channels. Using, for our example, a 256 channel ADC, we can now look back at the upper horizontal axis in each histogram in Figure 3.6. Having used the graph paper scale on the lower axis to plot intensity, we have now divided up the whole range of light intensity into 256 channels on the upper axis (numbered from 0 to 255); varying the voltage, offset, and gain simply assigns a different intensity to each of the channels (like changing the type of graph paper used or the scale used on that graph paper to suit the plotting of different types or ranges of data). The voltages and gains having been selected for a given experiment, the output data are then simply recorded by the cytometer electronics as light intensity on a scale of 0 to 255 (or 1,023). It is only by knowing something about the amplifier and voltage settings that you can know what relationship those values of 0 to 255 bear to one another.

For example, if we are using an ADC with 256 channels and have selected a log amplifier with a gain that gives us 2 decades full scale, then we would know that a signal appearing in channel 200 has been given off by a particle that is 10 times as bright as a particle giving a signal in channel 72 (on a 2 decade full scale amplifier, every 128 channels represents a 10-fold increase in intensity; the entire 256 channels represents two consecutive 10-fold increases in intensity = $10^2$). If, on the other hand, the gain on the log amplifier had been set to give us 4 log decades full scale, then a signal in channel 200 would come from a particle that is 100 times brighter than a particle giving a signal in channel 72 (on a 4 decade scale, every 64 channels represents a 10-fold increase in intensity; 128 channels represents a $10^2$-fold increase; the entire 256 channels represent a $10^4$-fold increase). If we had been using a linear amplifier, then a signal in channel 200 would represent a particle 2.78 times brighter than one with a signal in channel 72 (200/72 = 2.78). These examples should serve to illustrate that a knowledge

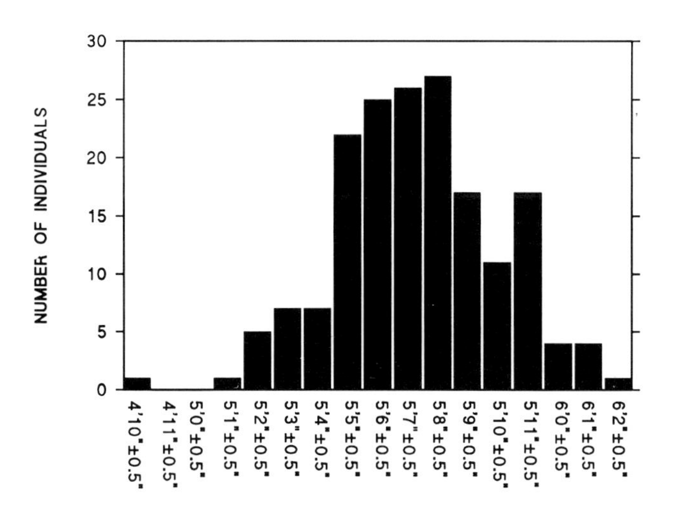


**Fig. 3.7.** World War One cadets from Connecticut Agricultural College arranged to form a height histogram. Photograph from A. Blakeslee (1914).

of instrument electronics and the settings used is required if we really want quantitative information about the relative brightness of signals from different particles; a simple channel number is not really enough.

One other capability we have with cytometry electronics is the definition of a *threshold.* An electronic threshold is just like a threshold into a room: It defines an obstacle. Only particles giving signals greater than that obstacle will be registered on the cytometer ADC. The most common use of this threshold is in the definition of a forward scatter channel number. Only particles with a forward scatter signal brighter than the defined channel threshold will be counted by the cytometer. By the use of a forward scatter threshold, we can avoid problems that might come from dust, debris, and electronic noise in the system. The dim forward scatter signals from debris and noise are not bright enough to pass over a threshold, and therefore "particles" of this type are completely ignored. There are other ways of using a threshold (for instance, by using a red fluorescence threshold to exclude from an ocean water sample the signals from any particles that do not contain chlorophyll), but the use of a forward scatter threshold is by far the most common.

To recapitulate, before we leave the discussion of electronics, we have now illuminated a particle; registered the light emerging from that particle onto one or another photodetector depending on the color or direction of that emerging light; and converted the signal from each photodetector to a measure of intensity on a scale of 0 to 255 (or 1,023). Therefore, for each particle that has flowed past the illuminating beam, we now have, simply, four or five or more numbers (depending on the number of photodetectors present) that describe that particle. Those four or five numbers (each on a scale of 0 to 255) tell us the intensity of the FSC, SSC, and fluorescence (red, green, orange) from that particle. And those numbers are, quite simply, the only information we now have about that particle. These are the facts that can be correlated with each other for data analysis. But that story is told in a later chapter. Before we get too far away from that particle in its stream, we want to back track a bit and discuss sorting.

## SORTING

Although flow cytometers were originally developed as instruments that could sort particles based on analysis of the signals coming from those particles, it is now true that most cytometers are used for analyzing particles but not for sorting them. This is just one of the many examples of the way in which flow cytometry technology moves in unpredictable ways. Many cytometers now do not even possess a sorting capability, and the instruments that can sort particles may be used only infrequently for that purpose. Nevertheless, sorting cells with a flow cytometer is an elegant technology; it may be the only method available for obtaining pure preparations of particles with certain kinds of characteristics of interest. On both of those counts, it is certainly worth knowing about even if you are not interested in sorting at the moment.

If a stream of liquid is vibrated along its axis, the stream will break up into drops. The characteristics of that drop formation are governed by an equation that will be familiar to anyone who has studied light waves: $v = f\lambda$, where v is the velocity of the stream; f is the frequency of the vibration applied; and $\lambda$ is the "wave length" or distance between the drops. One other fact that needs to be known before we can understand sorting is that drops form in a regular pattern so that the distance between drops is always equal to just about 4.5 times the diameter of the stream. For the usual flow cytometer, if a 70 μm nozzle is used, then the drops will form with a wave length of about 315 μm. In most cytometers, the stream velocity is approximately 10 m/sec; this means that we will have to vibrate the nozzle at

$$f = (10)/(315 \times 10^{-6}) = 31{,}746 \text{ cycles per second}$$

to get drops to form. And this, in turn, means that if the sample is of a concentration such that particles are moving down that stream at 30,000 particles per second, there will be one particle in every drop; if the particles are less concentrated and flowing at 3,000 particles per second, there will be a particle in every tenth drop.

If we observe a vibrating stream, we will see that drops break off from the jet at some measurable distance from the point of vibration. Therefore a particle can be illuminated and its signals detected as it flows within the core of a stream through a vibrating nozzle. At some distance after the point of analysis, the stream will begin to break up into drops, and the particles will be contained in those drops as they break off. It just remains now for the cytometer to put a charge on the drop containing a particle of interest; because the drops will flow past positively and negatively charged high-voltage plates, any drop carrying a positive or negative charge will be deflected out of the main stream and toward one or the other plate. Once this happens, it is easy to place containers in position to catch the drops deflected to the left or right. The main trick of flow sorting is therefore to apply a charge to just the right drop (*i.e.*, the one containing the desired particle). To do this, we need to know the time that elapses between the time a particle is analyzed at the analysis point and the time a bit later that that particle is just about to be trapped in a newly formed drop leaving the stream. If the entire stream is charged (either negatively or positively) by applying a charge at the nozzle just before the drop is formed with the desired particle within it, and then the entire stream is grounded at the nozzle to remove that charge just after the drop in question is formed and has detached from the stream, then only the drop with the desired particle within it will remain charged. What we need to do is determine the time it takes a particle to move from the analysis point to the point where the drop containing it breaks away from the stream.

Different methods are available for measuring the time delay between analyzing a particle and trapping that particle in a drop. In most cases, we simply measure the distance between the analysis point (the light spot on the stream) and the drop break off point; then, to convert that distance into a time, we count the number of drops that occur in an equal distance measured further down the stream. Since we can know the frequency of vibration that is driving the nozzle (in cycles per second), we therefore also know the number of seconds per cycle. By counting the number of drops that occur in a distance equal to the distance between analysis point and break off point, we can convert the counted number of drops into a time in seconds. And it is that time that is the time delay we want between analyzing the particle and charging the stream.

Fortunately, most cytometers will do most of these calculations for us. We simply need to count the number of drops that occur in a distance equivalent to the distance between analysis point and break off point. The cytometer electronics then convert that drop number into a time delay, and that time is the amount of time we need to delay our charging of the stream so that a particle analyzed and found to have desired characteristics has moved in the stream and is about to be trapped in a forming drop. Once the drop-delay time has been determined, the cytometer can then be given sorting windows delineating the flow cytometric characteristics of the desired particles: These windows are simply the numbers (again on a scale of 0–255) that describe the intensity of the light signals characterizing the particles of interest. If we are at all unsure about our accuracy in counting drops, then we can charge the stream for a slightly longer time (centered around the calculated time delay) so that three drops are charged after a desired particle has been detected at the analysis point. This should ensure that our desired particle is deflected, even if it should move slightly faster or slightly slower than anticipated. In fact, three drops are usually charged in most sorting applications (Fig. 3.8).

As discussed earlier, with a nozzle of 70 μm and a stream moving at 10 m/sec, our system is committed to a vibration frequency of about 30,000 cycles (30 kHz) per second. If we prefer a margin of safety, we probably want to charge and sort three drops at a time; in that case, we will want a particle in no more than every third drop. This means that our total particle flow rate can be no faster than 10,000 particles per second. To give us a bit more margin for error, most sorting operators like to have particles separated by about 10 empty drops. With a 70 μm nozzle and a stream velocity of 10 m/sec, this restricts our total particle flow rate to 3,000 particles per

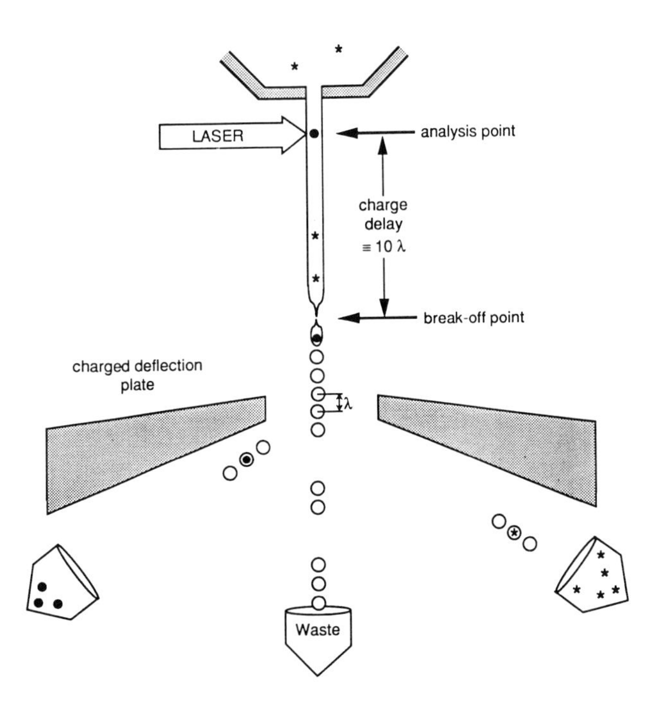


**Fig. 3.8.** Droplet formation for sorting. A charge delay is required between analysis of a particle and the charging of the stream so that only the drop (or three drops) surrounding the desired particle will be charged and then deflected.

second. For sorting cells of very low frequency within a mixed population, this may involve unacceptably long sorting times. The only way we can run our particles through at a faster rate is either by making the nozzle diameter smaller or by running the stream at a higher velocity. The first solution tends to be impractical, as a clogged nozzle ends up being very slow in the long run. The second solution has been implemented at Los Alamos and Livermore, where a high-pressure system brings about a stream velocity of about 100 m/sec; as a result particles flowing at rates of 30,000 particles per second can be sorted. But for most people with commercial cytometers, the total particle flow rate that can be used for sorting is restricted to about 2,000–3,000 particles per second in order to ensure that particles are no closer together than 1 in every 10 drops.

The implications of this particle flow rate can be seen from Table 3.2. If the rate of particles flowing is restricted to 2,000 per second, then the particles we actually desire to purify by the sorting procedure will be flowing at something less than that rate. If the desired particles are 50% of the total, then they will be flowing at a rate of 1,000 per second (3.6 million per hour). If, however, the desired particles are only 5% of the total, they will be flowing at a rate of 360,000 per hour. The time taken for sorting cells, then, depends, first, on what percent the desired particles are of the total number of particles present and, second, on how many desired particles are actually required at the end of the day. Scientists tend to think logarithmically: It is easy to say that perhaps $10^6$ or $10^7$ sorted particles are required for a given application. But sorting $10^7$ particles takes 10 times longer than sorting $10^6$ particles, and this can represent a very large invest-

**TABLE 3.2. Cell Sorting: Time Needed to Sort Required Number of Desired Particles at Total Particle Flow Rate of 2,000 Particles per Second**

| Required no. of desired particles | Desired particles as % of total particles | | | |
|---|---|---|---|---|
| | 0.1% | 1.0% | 5.0% | 50.0% |
| $10^3$ | 8 min | 48 sec | 10 sec | 1 sec |
| $10^4$ | 1.4 hr | 8 min | 1.7 min | 10 sec |
| $10^5$ | 14 hr | 1.4 hr | 17 min | 1.7 min |
| $10^6$ | 5.8 d | 14 hr | 2.8 hr | 17 min |
| $10^7$ | 1.9 mo | 5.8 d | 1.2 d | 2.8 hr |
| $10^8$ | 1.6 yr | 1.9 mo | 12 d | 1.2 d |

ment in time. The moral of this story is that any procedure that can be applied to enrich a particle suspension before flow cytometer sorting will save considerable amounts of time.

By way of a practical example that may be familiar to immunologists, lymphocytes are a class of mononuclear white blood cells that are divided into two types: B lymphocytes and T lymphocytes. There has been found to be a subclass of B lymphocytes (called $CD5^+$ B lymphocytes) that are of interest for functional analysis. Since $CD5^+$ B lymphocytes are 10% of all B lymphocytes, B lymphocytes are 10% of all lymphocytes, and lymphocytes are perhaps 50% of all mononuclear white cells, then, if we want to sort out $CD5^+$ B cells (0.5% of our total particles), they will be flowing through our system at a rate of 36,000 per hour if we use a preparation of mononuclear cells (with a total particle flow rate limited to about 2,000 per second). It would therefore take 28 hours to collect a million $CD5^+$ B cells. However, if we remove all nonlymphocytes by adherence, and then remove all T lymphocytes by rosetting (both techniques are rapid batch processes that might take an immunologist about 1 hour to perform), we can then send pure B lymphocytes through the cytometer, and the desired $CD5^+$ particles (now 10% of the total) will therefore be flowing at a rate of 720,000 per hour. As a result, the time it will take to get 1 million desired cells will go from 28 hours to 1.4 hours. This has obvious benefits in terms of both cost, if you are paying for cytometer time, and the general health and viability of the sorted cells at the end of the procedure.

This brings us to a brief discussion of the condition of cells after they have been sorted. Depending on the purpose to which they will be used, there may be different requirements for sterility and viability. Sterile sorting is possible; with care and a fair amount of 70% ethanol used to sterilize the flow lines, sorted cells can remain sterile and can be cultured for functional analysis. Given that being irradiated, vibrated, charged, and deflected must be slightly stressful, it seems reasonable to put the sorted cells into medium that will maintain their well-being as quickly as possible. If you will remember that the cells move down the stream in a core that is just a small percentage of the total volume of the sheath stream, you will realize that, following drop deflection, the cells are deposited in medium that is partly sample medium, but mainly sheath fluid. For this reason, the sheath fluid should be compatible with the cells in question.

We have now followed our particles from their sample container, down through the center of the sheath stream, past their intersection with the light beam, and (if sorting is in order) into a drop that is either charged or not.

Meanwhile, the signals from that cell have been registered on each of several photodetectors, and the intensity of those signals have been recorded as numbers (on a scale of 0 to 255 or 0 to 1023) in the temporary memory of a computer. What we now do with that numerical data is up to the computer software available.

## FURTHER READING

Chapter 2 in **Melamed et al.**, Chapters 4 and 6 in **Shapiro**, and Chapter 1 in **Ormerod** are good general descriptions of cytometer characteristics.

Chapter 3 in **Melamed et al.** and Chapter 3 in **Van Dilla et al.** discuss hydrodynamics and flow chamber design in depth.

Chapter 2 in **Watson** has a good discussion of fluid flow dynamics and of the avoidance of coincidence events.

# 4

# Information: Harnessing the Data

## DATA STORAGE

Having left the flow cytometry system with light signals from the sample particles recorded in appropriate channels on the ADC, we are now faced with the prospect of losing all these data as soon as we start recording data from our next sample (or as soon as we turn off the cytometer). What is obviously required is a way to store the data permanently for later correlation and analysis at our leisure. At this point we leave cytometry *sensu stricto* and find ourselves in the realm of computer ware (soft and hard). The main challenge encountered with storage of data from flow cytometry arises from the ability of a flow system to generate large amounts of data very quickly. In a four-parameter configuration (four photodetectors), each particle flowing past the light beam generates four signals (four bytes of information). If we want to analyze 10,000 cells for each sample (this sounds like a lot to someone used to microscopy, but flow cytometrists can analyze 10,000 cells in, say, 10 seconds, and therefore are easily persuaded that a large number of cells give statistically better information), then each sample that is run through the flow cytometer will generate 40,000 bytes (plus some extra for housekeeping arrangements). A six-parameter cytometer will generate somewhat more than 60,000 bytes from that same sample.

Therefore (staying with our downmarket four-parameter cytometer), the information from just 6 samples will fill a 250 Kb floppy disk; 12 samples will generate enough data to fill a 500 Kb disk. Floppy disks are readily available and may be the answer to data storage problems if the experiments are small; they will be an expensive and cumbersome answer if the experiments are large. Although all flow cytometrists start out with small experiments, most progress rapidly to experiments with 20 or 30 or more samples (think of using 96 well microtiter plates). In fact, one of the surest rules of flow cytometry is that data will, in less time than predicted, fill all available storage capacity. When you find yourself continually running out of floppy disks, you will begin to try to think of other options for data storage. The other options for flow cytometry data storage are just the same as the options for any other kind of computer data storage—generally hard disks, optical disks, and tapes (Table 4.1). Which options are available will be determined or limited by the particular software being used to drive the data storage, but a favored system is the use of a 20 Mb or 80 Mb hard disk, which will store the information from 500 or 2,000

**TABLE 4.1. The Capacities of Various Types of Storage Media**[a]

| | | Sample size (4-parameter list mode acquisition) | | |
|---|---|---|---|---|
| Media | Capacities | 5000 Cells (20 Kb) | 10,000 Cells (40 Kb) | 50,000 Cells (200 Kb) |
| Floppy disks | 250 Kb | 12[b] | 6 | 1 |
| | 750 Kb | 37 | 18 | 3 |
| | 1 Mb | 50 | 25 | 5 |
| Fixed disks | 20 Mb | 1,000 | 500 | 100 |
| | 80 Mb | 4,000 | 2,000 | 400 |
| Optical disks | 650 Mb | 32,500 | 16,250 | 3,250 |
| Tapes | 16 Mb | 800 | 400 | 80 |
| | 67 Mb | 3,350 | 1,675 | 335 |
| | 133 Mb | 6,650 | 3,325 | 665 |
| Removable rigid disks | 2.5 Mb | 125 | 62 | 12 |
| | 50 Mb | 2,500 | 1,250 | 250 |
| | 400 Mb | 20,000 | 10,000 | 2,000 |

[a]The capacities are recorded as the numbers of samples (of either 5,000, 10,000 or 50,000 cells each) whose list mode data after acquisition can be stored to media of the indicated size. The capacity in bytes of the different media are representative but will vary with the formats of different computing systems. Similarly, different acquisition software will require more or less extra storage space for the housekeeping information that is stored with each sample.
[b]Numbers of samples.

samples (10,000 cells each; four-parameter system). Hard disks have two main drawbacks. The first is that they are often "fixed": You cannot take them home with you, and one drawback of this is that someone else who has access to the system can wipe out your data (and, given enough time, probably will do just that). The second problem with hard disks is derived from that rule about data filling all available storage capacity: No matter how large the capacity of the hard disk, it will become full sooner than expected.

The solution to both of these problems is to have a back-up facility. This back-up facility could consist of removable (Bernoulli) hard disks or optical disks. The cheapest (and slowest) option is computer tape. With any luck, you will have backed your data from the fixed disk to the tape or removable disk and will have the data safely in your desk when someone, by mistake or because he or she needs more storage space, zeroes the system. For these reasons, a back-up facility of some type is a necessary requirement for multiuser flow cytometers.

## DATA ANALYSIS

Now that the data have been safely stored, we come to analysis, which is the real point of everything we have done so far. Methods for data analysis vary. They vary with the inclinations of the software programmer; they also vary with the budget of the cytometer facility. They may be strictly commercial or they may be homemade. These days, commercial manufacturers of cytometers compete with each other on the basis of their software systems as much as on the basis of their cytometer technology. In addition, independent entrepreneurs have recently begun to program for analysis of flow data and are successfully entering into the commercial market. Moreover, there are research groups like those at Los Alamos and Livermore who will supply software that they have written without charge. It is increasingly true that the software available for analysis plays a large role in what the user sees as an acceptable cytometry package. Samples may be run through a cytometer and the information from those particles stored very quickly; analysis and reanalysis of that information may then require a great deal of time. Therefore it is not surprising that software is an important aspect of flow cytometry.

In theory, data from all flow cytometry systems are now stored in so-called flow cytometry standard (FCS) format. This means that, although data stored after acquisition on one cytometer may not be analyzable on software from another cytometer (because manufacturers have been discouragingly slow at fully embracing the standard), the format is one that can be learned, and anyone with good programming skills could write software for analysis of flow cytometry data. The FCS standard also means that independent programmers can and do write programs that will handle data acquired on any cytometer. In practice, most people use commercial packages for data acquisition and storage because these packages are commonly purchased along with the cytometer. Although the same packages provide methods for data analysis, there are times when additional analysis software from an independent source may be helpful. This might be to provide advanced analysis methods, to provide especially pretty pictures for slides and publication, to store data to or analyze data from a data base, or to allow analysis on one or another home computing system.

The data stored in FCS format are usually "list mode" data. As described above, this means that, in a four-parameter cytometer, four bytes of information are stored for each particle. The entire data storage file consists of this four byte description of each particle in the sample. By retrieving the stored data, each particle can be analyzed again and the intensity of each of the four signals for that particle will be known and can be correlated with the intensity of the four signals from any other particle. This type of list mode data is useful because no cytometric information has been lost and it can all be examined again in future analyses. There are, however, other types of data storage systems that have the advantage of requiring less storage space. So-called single-parameter data storage involves the storage of the intensity profiles for the population of cells in a sample for each parameter separately; the only information stored is, for example, the distribution of forward scatter signal intensities for the cells in the sample; the distribution of side scatter signal intensities for the cells in the sample; the distribution of red fluorescence signal intensities for the cells in the sample; and the distribution of green fluorescence signal intensities for the cells in the sample. But in this case no information has been stored about whether the bright green particles are the particles that are bright red or whether they are the particles that are not red. With this kind of storage, we will not know whether the particles with a bright forward scatter signal are red or green or both red and green. Therefore, if we have stored data as single parameter information, we tie up very little storage capacity but severely restrict our options for future analysis. There is an in-between option called dual parameter correlated data, which allows correlation between two parameters and which uses up an intermediate amount of storage space. Unless storage capacity is very limited and the information required from data analysis is also very limited, list mode data storage is by far the best and indeed the only recommended option. Another rule about flow cytometry data analysis is that you are always going to want more information out of a sample than you thought was required when you planned and carried out the experiment. So use list mode data storage unless there is a very good reason for doing otherwise!

Having stored list mode data for all the particles in a sample, the software allows the correlation of the data in all possible directions. To understand the methods available for this analysis, we should define two terms: *gate* and *cursor* (or *marker*). A gate is a way of defining the characteristics of particles (in terms of FSC, SSC, and/or fluorescence intensity channel numbers) that we want to include in our analysis. Any particle that fulfills the defined characteristics will pass through the gate and will be included in the next analysis step. We have already used the concept of a gate in defining the characteristics of cells to be sorted. In addition to a "sorting gate," we can also use a "live gate" or an "analysis gate." A live gate defines the characteristics of cells that need to be fulfilled before the data from those cells are accepted for storage in the computer memory for further analysis; the information about all other cells will be lost. An analysis gate, by contrast, selects cells with defined characteristics from within a large heterogeneous sample that has already been stored in a data file. The use of either type of gate is rather analogous to the process of, by eye, disregarding all erythrocytes, monocytes, and polymorphs in a microscopic field and then counting only lymphocytes to determine the percentage of lymphocytes that are stained. A trained microscopist's eye is, in that way, defining a lymphocyte gate based on nuclear pattern, shape, and size. A cytometrist defines a lymphocyte gate in terms of FSC and SSC.

By way of illustration, we can go back to the example of $CD5^+$ B lymphocytes (as you might recall, the $CD5^+$ B lymphocytes are about 10% of all B lymphocytes, which are 10% of all lymphocytes, which in turn are about 50% of all mononuclear cells). If we want to analyze a peripheral blood mononuclear cell preparation (stained for a B-lymphocyte marker and also for the CD5 marker) to see what percentage of the B lymphocytes possess the $CD5^+$ phenotype, we could store the data from 10,000 cells and then use an analysis gate to restrict our analysis to lymphocytes; this would, however, give us too few $CD5^+$ B lymphocytes (perhaps 50) to provide a statistically accurate picture of the characteristics or frequency of these cells. We would need to store the data for about 200,000 cells in order to give us enough B lymphocytes (10,000) to analyze accurately those that are $CD5^+$ (perhaps about 1,000). This might be feasible,but would require a great deal of data storage space (800,000 bytes for list mode storage of four-parameter information from each sample). To save data storage space, we might choose to use a live gate—only storing the data from cells that are positive for the B-lymphocyte marker. In this way, we would use up only 40,000 bytes to permit analysis of the 1,000 or so lymphocytes that are $CD5^+$ of 10,000 B lymphocytes. However, with this live gate we have lost some information that might actually be of interest. Our data file for this sample will not, for example, allow us to calculate the percentage of lymphocytes that are B lymphocytes because the live gate has excluded from storage all the non-B lymphocytes. Only with the data stored without a live gate could we calculate this value.

Once a gate has been used to define the particles of interest, we then have several options available for analyzing the characteristics of those particles. We can readily look at any one parameter in isolation and analyze the light intensity histogram produced by the particles from a sample with respect to that parameter. Since we have stored the channel numbers characterizing the signals from each cell, the software can plot a histogram (number of particles at each channel as in the height histogram; see Fig. 3.7), with all the cells placed according to the channel number characterizing the intensity of their signals. In this way we could look at the intensity distribution of, for example, green fluorescence signals from 10,000 particles from one sample. And then we could look at the intensity distribution of red fluorescence signals from the same 10,000 particles. We can, in fact, generate a histogram for each of the parameters measured (Fig. 4.1). Some software will plot data according to channel number; other software will convert the channel data and use a "relative intensity" scale (think of the upper and lower horizontal axes in Fig. 3.6). In the latter case, the software is making assumptions about the accuracy, linearity, and

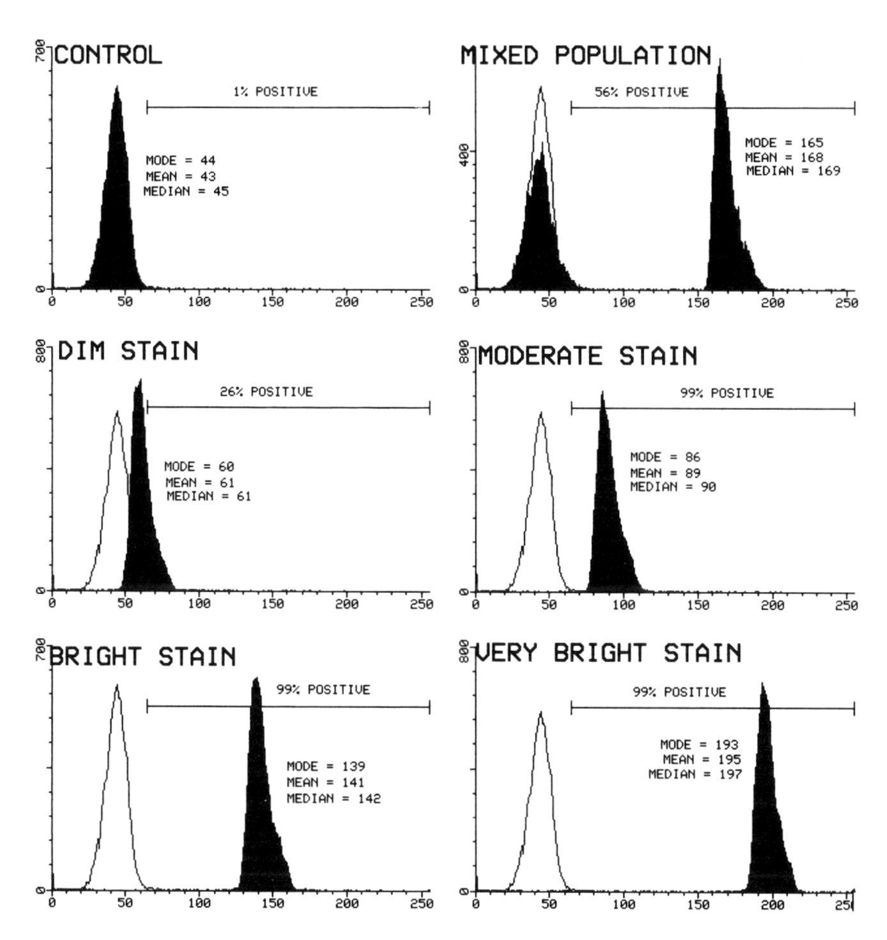


**Fig. 4.1.** Methods for describing the histogram distribution of signal intensities from a population of particles. The plots show the number of particles on the vertical axis against channel numbers on the horizontal axis. If cursors or markers are placed to delineate a region of positive intensity (relative to the 1% level on an unstained control), the "% positive" value can usefully describe a mixed population of stained and unstained particles. This value will be misleading if used to describe a uniform population of dimly stained cells. The "mode," "mean," or "median" channel number is the best way to compare uniform populations of cells of varying fluorescence intensity.

amplification gain on the photodetectors. Once we have plotted the histogram distribution (number of particles at each defined light signal intensity), the software will allow us to analyze this distribution to extract certain kinds of information: what percentage of the cells fall within a specified intensity range, what the most common intensity (channel number) is for the group of cells (the "mode" channel), what the mean intensity channel is for the group of cells, and what the median intensity is for that group.

Just how these values are obtained will vary with the particular software in question. "Cursors" or "markers" can be used to define regions of intensity that may be of interest; for example, we could place a cursor so that it separates the low intensity range of green fluorescence from the high intensity range of green fluorescence, and we could then ask how many particles fall within the high intensity range. A value for the percentage of positively stained cells can be determined by placing a cursor at a position defined by the background fluorescence of unstained cells; by convention at the 1%–2% level, this kind of cursor is usually the best way to describe a mixed population that consists of both unstained cells and brightly stained cells.

If we are, on the other hand, concerned with the changing fluorescence intensity of a uniform population of cells, we would be better served by using mode or median or mean characteristics of that population (the use of a "percent positive" value is, in fact, highly misleading if we are looking at a population of cells that are uniformly but dimly fluorescent). The mode value, being simply the channel number describing the most common group of cells, may vary erratically and be poorly reproducible if the population distribution is very broad. The mean value will be incorrect if significant numbers of cells are in the highest or lowest channels (0 or 255) of the histogram. The median value for the population is most reproducible, but is only applicable if all the cells in a population are to be analyzed (Fig. 4.1).

If we now want to go further and correlate one parameter with another, most software analysis packages implement the plotting of so-called *dot plots* (sometimes called *bit maps*) or *contour plots*. These plots allow the correlation of any two parameters with each other: Each particle is placed on the plot according to its intensity channel for each of the selected two parameters. Dot plots show, simply, a dot on the page or screen at each locus defining quantitatively (according to channel number) the two characteristics of each particle in the sample. Six two-parameter correlation plots can be derived from our four-parameter data (Fig. 4.2). Dot plots suffer, graphically, from black-out (or green-out on a computer screen) in that an area of a display can get no darker than completely black (or green); if the number of particles at a given point are very dense, their visual impact, in comparison with less dense areas, will decrease as greater numbers of particles are displayed. Contour plots, on the other hand, display the same kind of correlation as dot plots, but can provide more visual information about the density of particles at any given point in the correlation display. They allow the display of data according to the number of cells (density) in any particular region (think of the contour lines describing peaks on a mountaineer's contour map). Lines are assigned to various levels of cell density (as contour maps assign lines to different altitudes) according to any one of several different strategies. While changes in the assignment of lines will not change the values calculated for the number of cells in a given region or quadrant, they may radically change the way the data set appears. Figure 4.3 shows examples of how the same data plotted with different cell density assignments for contour lines can make messy data look tidy (or vice versa) and can even make three peaks look like a single distribution. The message here is simply that we need an informed eye when looking at contour plots generated from other people's data.

Once the data from a sample have been plotted in two dimensions, the distribution of particles can then be analyzed. *Regions* of irregular shape or *quadrants* dividing the plot into four rectangles are terms that are used for the cursors applied in two-dimensional analysis. As with histograms, these cursors simply divide the light intensity channels into areas of interest. The number of particles in each of the defined areas can then be counted (Fig. 4.4). A standard analysis procedure might use unstained control cells to define the channels delimiting background red and green fluorescence. Quadrants can then be drawn based on these channels, and the quadrants will therefore define the staining intensities that we would consider to represent green particles, red particles, and so-called double-positive particles that are both red and green, as well as the double-negative (unstained) region.

As an intellectual exercise, the experimenter should always attempt to draw mentally the single dimensional histograms that would result from contour plot data. Figure 4.5 indicates two different contour plots (with

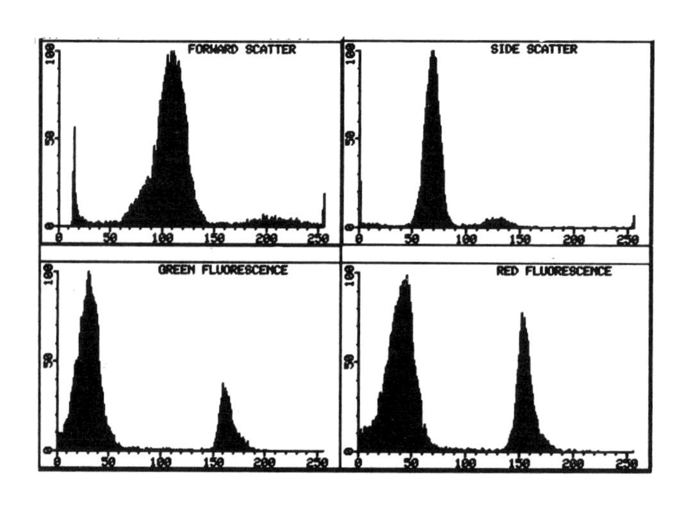
FORWARD SCATTER
SIDE SCATTER
GREEN FLUORESCENCE
RED FLUORESCENCE

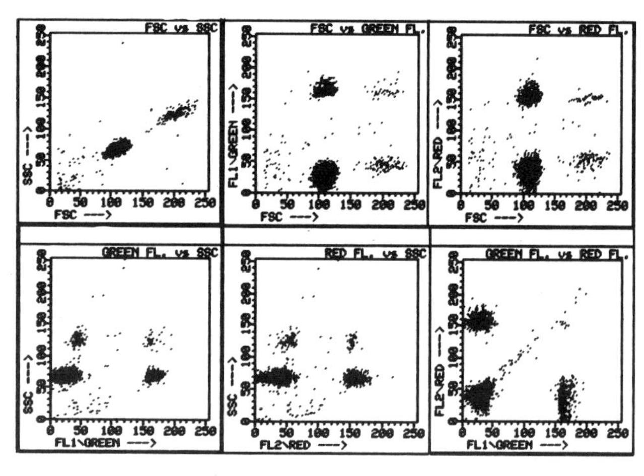
FSC vs SSC
FSC vs GREEN FL.
FSC vs RED FL.
GREEN FL. vs SSC
RED FL. vs SSC
GREEN FL. vs RED FL.
SSC --->
FSC --->
FL1\GREEN --->
FL2\RED --->

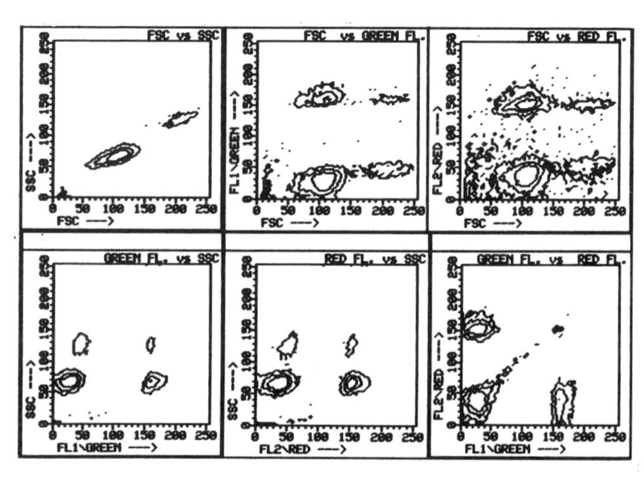
FSC vs SSC
FSC vs GREEN FL.
FSC vs RED FL.
GREEN FL. vs SSC
RED FL. vs SSC
GREEN FL. vs RED FL.
SSC --->
FSC --->
FL1\GREEN --->
FL2\RED --->

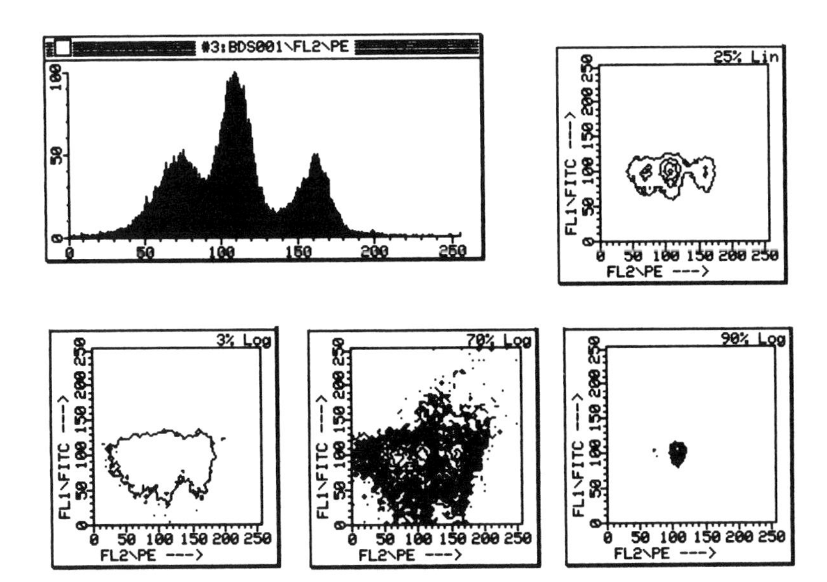


**Fig. 4.3.** A histogram of fluorescence and contour plots (plotted according to different line assignments) of the same data. Comparison between the histogram and the contour plots can allow us to see at what "altitudes" the contour lines have been drawn for each contour plot and why the resulting displays look so different.

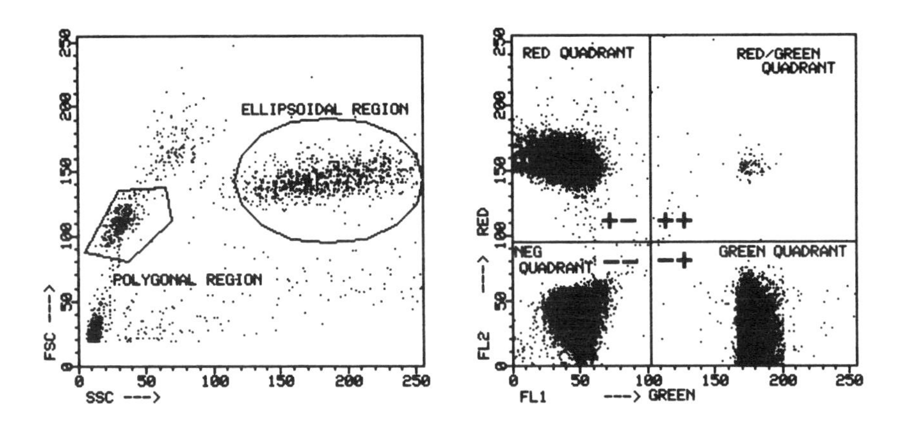


**Fig. 4.4.** Regions and quadrants are methods for delineating the intensity (according to channel number) of cells of interest in dual-parameter analysis.

---

**Fig. 4.2.** The four histograms and six dual-parameter plots derived from the data from a four-parameter cytometer. Both dot plots and contour plots are shown as alternate methods for displaying dual-parameter correlations.

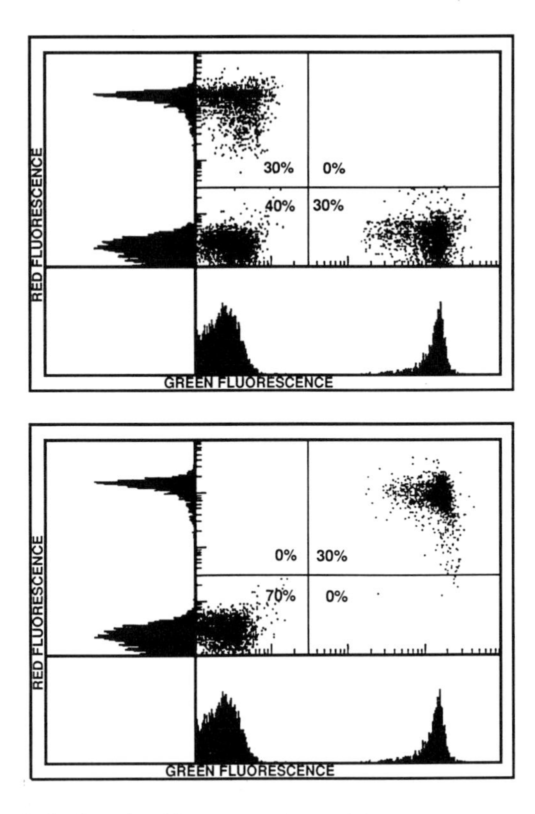


**Fig. 4.5.** Identical single-color histograms derived from quite different cell samples. The dual-parameter dot plots allow us to distinguish these two populations of cells.

their respective histograms); it is clear that the histograms from these two different sets of data are identical but that the dot plots are quite different. This example should serve to emphasize that more information is obtained from dual-stained preparations with their two-dimensional plots than from two successive single-dimensional (one-color) analyses.

Dual-parameter correlations constitute the standard procedure for analysis of most flow cytometric data. Most cytometers provide us with four or five or more parameters of information. Because humans, in general, are used to thinking in two dimensions, actually correlating three or more parameters with each other can seem rather difficult—in terms of both computer software and our ability to keep track of the strategy (Fig. 4.6). For purposes of data analysis, the most common procedures use the extra parameters to allow gating (including or excluding) of particles before the final analysis, which is almost invariably just two dimensional.

By now, we should have a reasonably clear picture of the physical and electronic characteristics that form the basis for flow cytometric analysis. We have followed particles into the center of a stream flowing through a nozzle into a light path; we have seen how reflection, refraction, and fluorescence can generate light signals from those particles as they are hit by that light beam; we have accounted for the registering of those signals by photodetectors so that their intensities can be assigned to the discrete channels (256 or 1024) of an ADC; we have described the way in which the voltage and amplification applied to the photodetector signals will determine the intensity range represented by the ADC channels and their relationship to each other in terms of absolute intensity; we have seen the way in which the channel number for each of the parameters characterizing

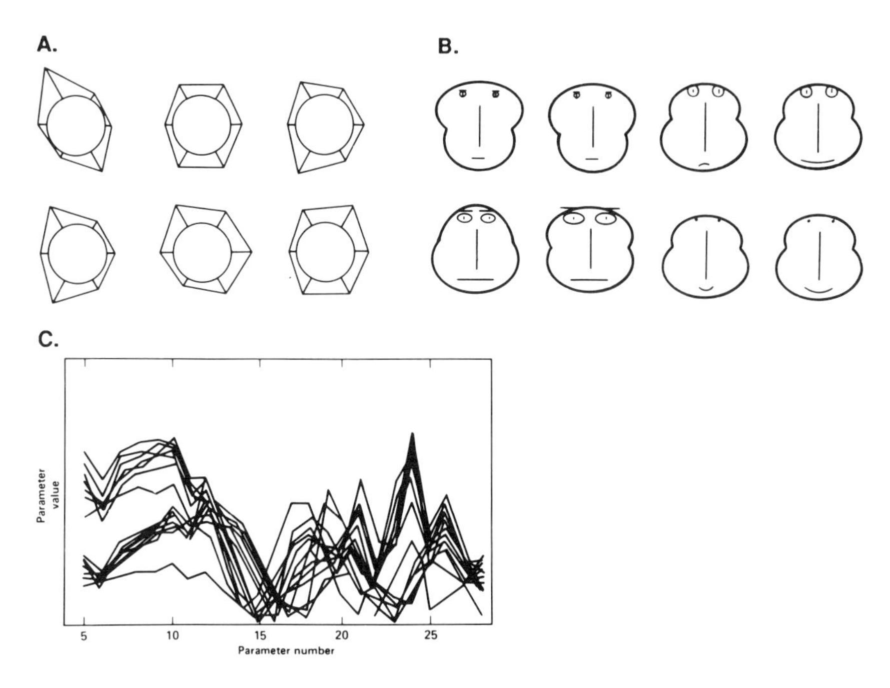


**Fig. 4.6.** Three different ways to visualize multivariate data. From Dean (1990).

each particle can be stored as list mode data for each sample; and we have described the general ways in which the stored data can be displayed and analyzed. The chapters that follow will deal with the ways in which these principles are applied to experimental situations. Because the first decisions that are made in designing a flow experiment often concern the specification of reagents for staining the particles in question, we will initially take a short detour back to the principles of photochemistry and laser optics in order to understand the requirements that flow cytometry imposes on these decisions.

## FURTHER READING

Chapter 5 in **Shapiro**, Chapter 30 in Volume I of **Weir**, and Chapter 22 in **Melamed et al.** are all good discussions of some of the mathematical aspects of flow data analysis.

# 5

# Seeing the Light: Lasers, Fluorochromes, and Filters

Because flow cytometry involves the illumination of particles by a light source and usually involves subsequent analysis of the fluorescent light emitted by fluorochrome-stained particles after illumination, an understanding of some of the principles behind light production, light absorption, and light emission is important for the effective design of experimental protocols. The elementary concepts of photochemistry apply both to the generation of light by a light source and to the absorption and emission of light by a fluorochrome; an understanding of these basic concepts can help us to avoid hazardous pitfalls.

## GENERAL THEORY

It is worthwhile to begin with a brief review of atomic structure. Atoms consist of relatively compact nuclei containing protons and neutrons. At some distance from these dense nuclei each atom has electrons moving in a cloud around the central nucleus. The electrons move in shells or orbitals or probability waves (different words derive from more or less classical or quantum mechanical terms of reference) around the nucleus, and the number of electrons circulating in these orbitals depends on the element in question. Four things are particularly important for flow cytometrists to understand about these electrons: First, atoms have precisely defined orbitals in which electrons may reside. Second, an electron can reside in any one of the defined orbitals but cannot reside in a region that falls between defined orbitals. Third, the energy of an electron is related to the orbital in which it resides at any given moment. Fourth, an electron can absorb energy and be pushed to an excited (higher energy) orbital, but it will quickly give back that energy as it rapidly returns to its stable, ground-state configuration (Fig. 5.1).

Having summed up all of electronic structure with four facts, we can now add to our knowledge one more fact about light itself. Light is a form of energy made up of photons and the color of the light is related to the amount of energy in the photons of that light. For example, when red light shines on an object, that object is bombarded with photons of relatively low energy; blue light, by contrast, is made up of photons of higher energy. *Wavelength* is a term often used to describe the color of light, and the wavelength is inversely related to the amount of energy in the photons of that light: Blue light has wavelengths of about 400–500 nm and its photons have relatively large amounts of energy; red light has wavelengths of about 600–650 nm and photons with less energy than those in blue light. The other colors of the visible spectrum fall between those values (see Fig. 5.2).

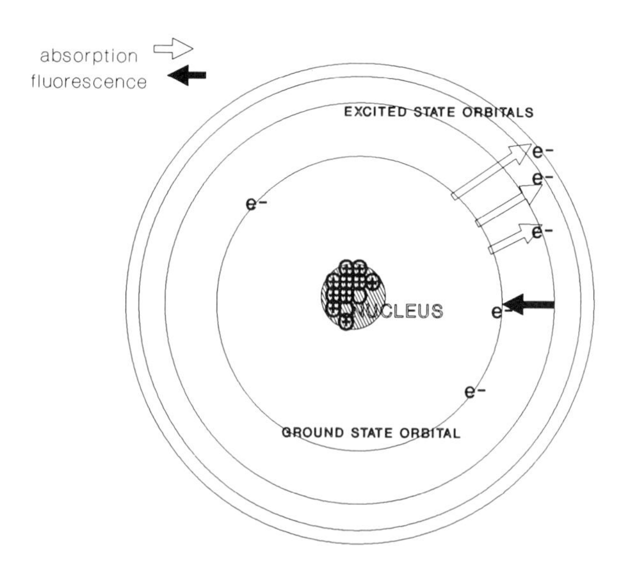


**Fig. 5.1.** An "old-fashioned" but conceptually easy diagram of an atom with electrons circling the nucleus. Electrons can absorb energy to raise them to an excited state orbital. When they fall back to their ground state, they may emit light, which we call *fluorescence.*

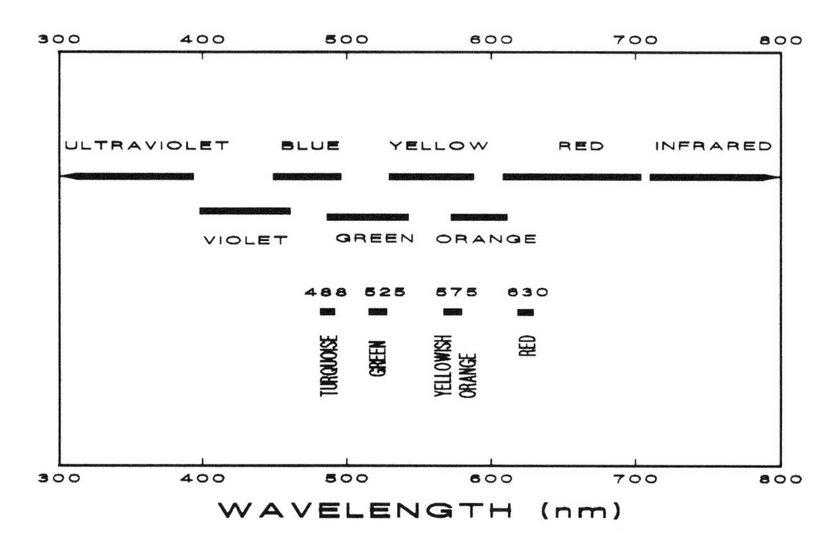


**Fig. 5.2.** Light, in the region to which our eyes respond, can be described by a "color" or, more precisely, by a wavelength. Photons with greater wavelength have less energy.

Infrared light has a longer wavelength than red light (and less energetic photons); ultraviolet light has a shorter wavelength than blue light (and more energy). With this information, we can now begin to understand a bit more about the flow cytometric system.

## THE ILLUMINATING BEAM

The illumination of particles as they flow past a light source is responsible for generation of the scatter and fluorescent signals upon which flow cytometric analysis is based. The illumination of particles can be provided by an ordinary arc lamp or by a laser. In either case, electrons within the light source are raised to excited orbitals by the use of electricity, and energy is given off in the form of photons of light when the electrons fall back to their ground-state orbitals. The color of that light is determined by the energy differences between the excited and ground-state orbitals of the atoms in the lamp or laser. Lasers and arc lamps each have certain advantages. Because lasers are the light source of choice in most of the currently used flow systems and because the use of a laser results in definite restrictions on experimental flexibility, the way in which a laser works is considered first, and then it will be compared, for better and for worse, with the traditional arc lamp.

Gas lasers consist of tubes (called *plasma tubes*) filled with gas; a cathode lies at one end and an anode at the other (Fig. 5.3). A voltage is applied across the plasma tube in order to raise the electrons in the atoms of gas to excited orbitals. As the electrons fall back to their ground state, they give off energy in the form of photons of light; the color of the emitted light is determined by the type of gas used and is a function of the energy levels of its orbitals. Reflecting mirrors are placed outside the tube at either end. If the plasma tube is aligned precisely with the reflecting mirrors, then the photons given off by the gas will be reflected back and forth through the tube between the two reflecting mirrors. The applied voltage maintains the electrons in the gas in excited orbitals. The reflected photons, as they oscillate back and forth within the tube, interact with these excited gas ions to stimulate them to release more photons of the same energy and phase (in a way predicted by Einstein). An amplification system thereby results: The light oscillating back and forth between the two end mirrors causes more and more light to join the beam. By allowing a small percentage of this oscillating beam to leave the system at the front mirror, we have generated what is known as a *coherent* light source.

A laser light source is coherent in three respects. It is coherent as regards its direction, in that a beam is generated that diverges little and remains

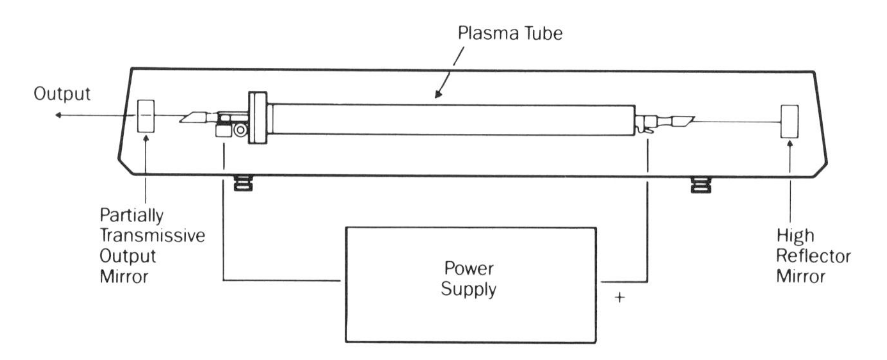


**Fig. 5.3.** The structure of a basic gas ion laser. Reproduced with permission of Spectra Physics.

compact and bright for great distances (as in laser light shows and "star wars"). It is coherent with respect to polarization plane (of possible relevance for some specialized aspects of cytometry). Lastly, it is coherent with respect to color (because the electrons in a gas atom or ion are restricted in the orbitals that they use under plasma tube conditions and are therefore restricted in the amount of energy that is emitted when each photon falls from an excited state to the ground state).

Coherence with respect to direction is the reason lasers are useful in flow cytometry: They provide a very bright, narrow beam of light allowing particles flowing in a stream to be illuminated strongly for a very short period of time so that measurable signals from one particle can be generated and then separated in time from signals generated by the following particle (see Fig. 3.5). The disadvantage of this spatial coherence is that, as was discussed in Chapter 3 regarding fluidics, cells in a stream must be well aligned in the center of that stream if they are going to be uniformly illuminated.

Although spectral purity has certain clear advantages, coherence with respect to color is actually one of the major limitations of a laser system. Whereas an ordinary light bulb will put out light of a wide range of colors (a white light bulb puts out a mixture of the whole range of colors in the visible spectrum), the color of the output from a laser is restricted by the restricted range of electron orbitals that will support lasing in any particular gas. Argon ion lasers are the most common lasers used in flow cytometry today; they put out useful amounts of light at 488 nm (turquoise) and at 514 nm (green), as well as small amounts of ultraviolet light; by using a prism, the operator or manufacturer can select one or the other of these colors, but there is relatively little light available in other regions of the spectrum (Fig. 5.4). Helium-neon lasers are also used (often in conjunction with an argon ion laser in a dual laser flow system); a helium-neon laser puts out red light at 633 nm.

A laser system is therefore very good at providing bright, narrow, stable light beams of well-defined color. It is inflexible, however, in terms of the color of that light. In addition, large lasers are demanding in their requirement for routine, skilled maintenance; careful alignment between plasma tube, reflecting mirrors, output beam, and stream is essential. And lasers are expensive; plasma tubes have limited lives and cost considerably more than an arc lamp when they need replacing. (It is worth noting that a well-aligned plasma tube with clean mirrors will last longer than a poorly aligned and dirty one because it will draw less current to produce the same amount of

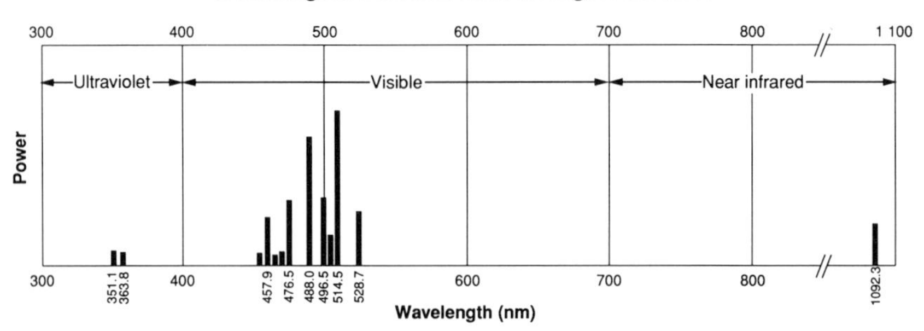


**Fig. 5.4.** The wavelengths of light emitted by an argon ion laser. The most powerful (and therefore most useful) wavelengths are 488 and 514 nm. High-power argon ion lasers can provide some useful light in the UV range.

light; loving care of a laser (and no smoking in the lab) can make a considerable difference to the overall running costs of a flow cytometer.) Some of the flow systems commercially available for analysis (without sorting capability) use small, air-cooled lasers. Air cooling is feasible when the laser is of low power and does not generate much heat (lasers are all relatively inefficient and generate a good deal of waste heat). However, cytometers with a stream-in-air analysis point have space between the particles and the light collecting lenses; as a result, detection of the light signals generated by the cells is inefficient. For this reason, sorting cytometers may require large, high-power lasers that put out very bright beams of light but also generate a great deal of waste heat. These sorting cytometers usually need a large supply of cold water to keep their lasers cool and also need skilled cleaning and alignment. The main fact to be kept in mind about lasers, as we move to a discussion of fluorochromes, is that, because lasers make use of electrons excited to a limited range of orbitals, they are restricted in the color of the light they emit.

## FLUOROCHROMES

In a laser, electrons are excited by means of energy from an applied voltage. We then take advantage of the energy given off as light by the electrons when they fall back to the ground state. With a fluorochrome, light itself (called the *excitation light*) is used to excite the electrons initially. We then analyze the emitted light (of a different color) that is given off as the electrons of the fluorochrome return to their ground-state orbitals. On the basis of our knowledge about electron orbitals together with our knowledge about laser output, it should now be apparent that, when light is shone at a fluorochrome, the electrons in that fluorochrome can absorb energy if, and only if, the excitation light contains photons with just the right amount of energy to push an electron from one defined fluorochrome orbital to a higher (more energetic) defined orbital. Thus the difference between orbital energy levels of a fluorochrome will strictly determine what color of light that particular fluorochrome can absorb. When electrons fall back from their excited (more energetic) level to their ordinary (ground-state) orbital, light is emitted that is of a color that depends on the difference in energy levels between the two orbitals in question. Because energy is not created and is, in fact, always lost to some extent from the system as heat, the color of the light emitted when an electron falls back from an excited state to its ground state is always of a somewhat lower energy and longer wavelength than the energy absorbed in raising the electron to the higher orbital in the first place (Fig. 5.5). The light given off as an electron falls back from an excited state to its ground state orbital is called *fluorescence.* The time required for fluorescence to take place is approximately 10 nanoseconds after the intial activation of the fluorochrome by the excitation beam.

If we plot the colors of light that can be absorbed and the colors that are then emitted (called *absorption* and *emission spectra*) by various compounds, we can see how these principles work in practice (Fig. 5.6). The wavelengths of light absorbed by a compound will depend exactly on the electronic orbitals of its constituent atoms; the light then given off as fluorescence is always of a longer wavelength than the absorbed light. The difference between the peak wavelength for absorption and the peak wavelength for fluorescence emission is known as the *Stokes shift.* Some compounds (for example, dansylchloride) have a larger Stokes shift than others (for example, rhodamine 101).

We are now in a position to understand why the use of a laser to provide the illuminating beam in a flow system restricts the choice of fluorochromes that can be used for staining cells. If we are using an argon ion laser with an output at 488 nm, we can consider as suitable stains those and only those fluorochromes that absorb light at 488 nm. Rhodamine, a stain used extensively by microscopists, absorbs light poorly at 488 nm and is therefore not useful in conjunction with an argon laser. Stains like DAPI

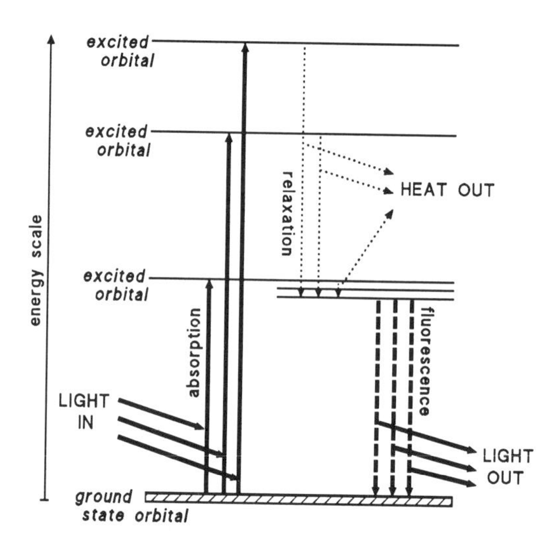


**Fig. 5.5.** The absorption of light by an electron and the subsequent emission of the energy from that light as both light and heat. Because some of the absorbed energy is lost as heat, the emitted light always has less energy (and is of longer wavelength) than the absorbed light.

and the Hoechst stains can be used with a high-energy argon ion laser tuned to the ultraviolet line, but cannot be used if the laser is tuned to 488 nm. Appropriate stains for 488 nm excitation include oxycarbocyanines for looking at membrane potential and propidium iodide and acridine orange for looking at nucleic acid content. With the 488 nm light from an argon laser, the situation also is ideal for staining cells with fluorescein (another standby of microscopists). In fact, this traditional allegiance to fluorescein (FITC) is the principal reason that argon ion lasers were initially selected for the first laser-based flow cytometers. Fluorescein absorbs light in the range of 460–510 nm and then fluoresces in the range of 510–560 nm, with a peak at about 525 nm (green); it can also be readily conjugated to antibodies, thereby providing specific fluorescent probes for cell antigens (Table 5.1).

A flow cytometer, however, has the ability to measure two or more fluorescence signals simultaneously on two or more photodetectors. To use several fluorochromes at the same time, cytometrists require a group of stains, all of which absorb 488 nm light but which have different Stokes

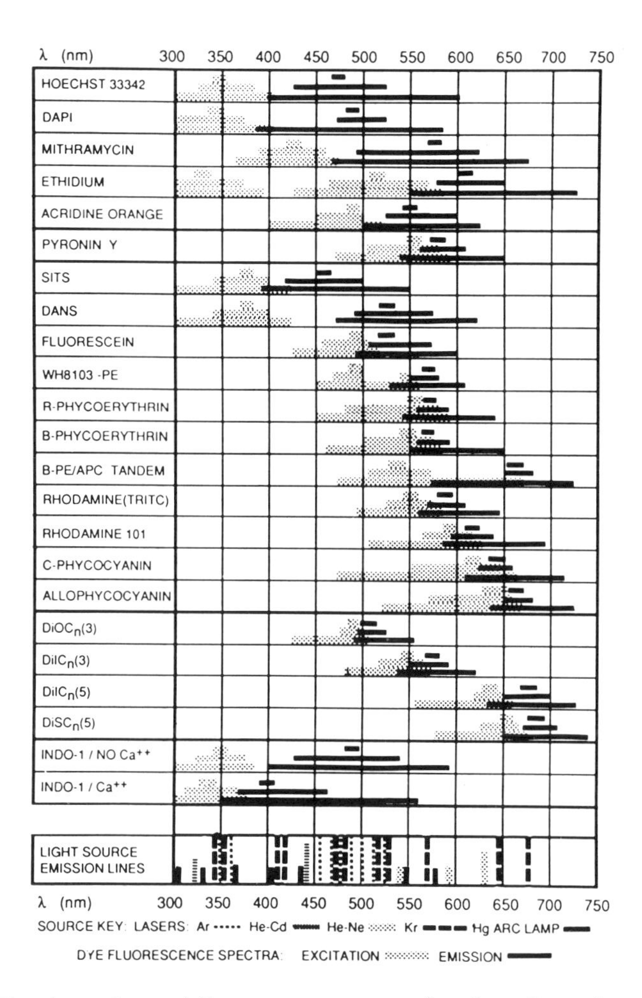


**Fig. 5.6.** The absorption and fluorescence spectra of various fluorochromes. From Shapiro (1988).

shifts so that they emit fluorescent light at different wavelengths and therefore will be detected on different photodetectors. Propidium iodide is one fluorochrome that fulfills these criteria (having a longer Stokes shift than fluorescein) and is, in fact, very useful for analyzing nucleic acid content in fluorescein-labeled cells. Propidium iodide, however, is not fluorescent unless bound to double-stranded nucleic acid and therefore is not usefully bound to antibodies for the staining of cell surfaces. Flow cytometrists were therefore spurred to hunt for appropriate pigments (absorbing 488 nm light,

**TABLE 5.1. Absorption and Emission Wavelength Maxima of Some Useful Fluorochromes**

| Fluorochrome | Absorption (nm) | Emission (nm) |
|---|---|---|
| Covalent labeling of proteins | | |
| Fluorescein (FITC) | 490 | 514 |
| R-phycoerythrin | 480–565 | 578 |
| Tetramethyl Rhodamine (TRITC) | 540 | 570 |
| Texas Red | 596 | 615 |
| Phycocyanin | 620 | 650 |
| Allophycocyanin | 650 | 660 |
| Nucleic acids | | |
| Hoechst 33342 or 33258 | 343 | 482 |
| DAPI | 345 | 455 |
| Ethidium bromide | 482 | 616 |
| Propidium iodide | 493 | 630 |
| Acridine Orange | | |
| + DNA | 480 | 510 |
| + RNA | 440–470 | 650 |
| Thiazole Orange | 509 | 533 |
| Membrane potential | | |
| di-O-$C_n$(3)-oxycarbocyanines | 485 | 505 |
| Rhodamine 123 | 505 | 534 |
| pH | | |
| Carboxyfluorescein | | |
| High pH | 495 | 520 (bright) |
| Low pH | 450 | 520 (weak) |
| BCECF | | |
| High pH | 508 | 531 (bright) |
| Low pH | 460–490 | 531 (weak) |
| Calcium | | |
| Indo-1 | | |
| Low calcium | 361 | 479 |
| High calcium | 330 | 405 |
| Fluo-3 | | |
| High calcium | 490 | 530 (bright) |
| Low calcium | 490 | 530 (weak) |

fluorescing at a wavelength different from 525 nm, and capable of chemical conjugation to proteins), and the humble seaweeds provided the answer.

Algae need to absorb light for photosynthetic metabolism, but a number of them are in the habit of living in dimly lit regions beneath the surface of the ocean. In this inhospitable marine environment, they have evolved a rich assortment of pigments. Surveying this array of algal pigments, we can find compounds that absorb light and fluoresce at a range of different wavelengths. A quick inspection of the absorption and fluorescence emission spectra indicates that, with a light source of 488 nm, phycoerythrin will absorb light fairly efficiently. With its longish Stokes shift, it fluoresces around 575 nm (orange). In addition, phycoerythrin was found to be almost as suitable as fluorescein for conjugating to antibodies or other proteins. For these reasons, fluorescein and phycoerythrin (PE) have become the fluorochromes of choice for flow cytometry. They can both be conjugated to antibodies to provide specific reagents that fluoresce with different colors.

To provide a third fluorochrome that absorbs light at 488 nm, that fluoresces at a wavelength even longer than 575 nm, and that can be conjugated to proteins, organic chemists have engineered "tandem" fluorochromes to fulfill this function in systems requiring the use of three simultaneous stains. They consist of a phycoerythrin molecule covalently linked to a Texas Red molecule or to an allophycocyanin molecule; in the presence of 488 nm light, the phycoerythrin moiety will absorb light, but will pass this energy on to the closely linked Texas Red/allophycocyanin molecule (by a process called nonradiative transfer, *i.e.*, no light is emitted), which will then fluoresce at its own fluorescence wavelength well beyond 600 nm. At the present time, this tandem molecule may be the best option for use with a 488 nm laser beam in a three fluorescence parameter system. However, a form of chlorophyll that absorbs 488 nm light, that has a long Stokes shift with a fluorescence maximum at 675 nm, and that can be easily conjugated to proteins shows promise for the future.

To increase their options within the limitations of the available fluorochromes, cytometrists are, at the present time, forced to purchase a second laser. By the use of a helium-neon laser, emitting light at 633 nm, algal pigments such as allophycocyanin can be used. A second argon laser also provides the option of using ultraviolet light at the same time as 488 nm light, and this obviously increases the range of fluorochromes that can be analyzed. However, extra lasers are expensive and are not an option that is available to many laboratories. Research cytometers can often be adapted for the implementation of two-laser illumination. But most routine, clinical cytometers do not have room for a second laser on their optical benches.

## LENSES, FILTERS, AND MIRRORS

We have now described a system in which a narrow beam of light of a well-defined wavelength is used to illuminate a particle, and the light scattered or fluorescing from that particle provides signals that are registered on a group of photodetectors. There are a few more optical elements in the system that help to direct the light in required directions. These elements are lenses, which focus light; filters, which selectively transmit light of particular wavelengths; and mirrors, which split light into component parts.

The lenses in a flow system are most importantly used to determine the precise cross-sectional profile of the illuminating beam. Although a laser beam is, by itself, narrow and parallel, it will have a cross-sectional diameter of several millimeters. A lens, placed in the laser path before the stream, may be able to improve on this a bit; a lens is usually used to focus the laser beam exactly at the stream with a cross-sectional diameter of about 20–50 μm. We want our particles to pass in and out of the beam as quickly as possible so that, even with a high particle concentration, we can resolve signals from two particles following closely to each other in the stream; but a narrow beam diameter may result in non-uniform illumination if the particles are not precisely maintained in the stream core. An illuminating beam with a circular cross section may therefore not be ideal. In fact, lenses can be used to provide illuminating beams with elliptical cross sections; these have the advantage of a narrow axis in the direction of particle flow but a wide and more tolerant axis across the stream width (Fig. 5.7).

Having illuminated our particles, we now come to the matter of making sure that we can send the different colored light signals coming from those particles toward different photodetectors. From our description of the optical bench in Chapter 3, we should recall that there are three, four, or more photomultiplier tubes whose job it is to measure the intensity of the light coming out at right angles from the particles in the illuminating beam. The light coming out at right angles will be a mixture of light of the same color as the laser beam (side scatter) and light of an assortment of other colors, depending on the fluorescence stains associated with the particle. It is only by restricting each photodetector to a response to a particular color of light that we can associate the signal from each detector to light from a

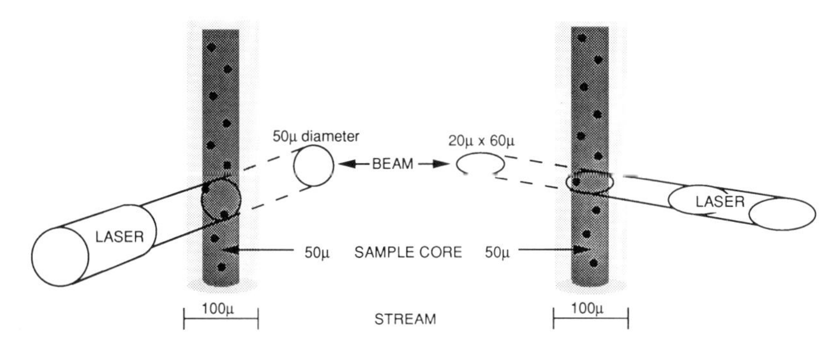


**Fig. 5.7.** An elliptical beam profile at the analysis point may allow the separation of signals from two closely spaced particles, but at the same time illuminate uniformly even those particles at the edge of a rather wide sample core.

particular fluorochrome and thereby obtain any specific knowledge about the staining or scatter characteristics of our particles. We need to use a combination of filters and mirrors to direct light of specific color toward its intended photodetector and not to any other (Fig. 5.8). Specialized mirrors can be used for this function. These mirrors are only partially coated with silver so that they split a light beam by reflecting some of the light shining onto their surface while transmitting the rest. One type of mirror used in the optics of flow systems is a mirror that splits light simply by percentage: It will, for example, transmit 95% of the light falling onto its surface and reflect 5%. This type of mirror may be used to split the side scatter (488 nm) light from the fluorescent component—by virtue of the fact that side scatter light is a great deal brighter than light fluorescing from stained particles and a photodetector can detect side scatter light even if only 5% of that scattered impinges on the tube surface. For analysis of the fluorescence from the particles, a different kind of mirror is required. These mirrors are called *dichroic mirrors*; they split a light beam not according to quantity but according to color. A 560 nm dichroic mirror, for example, will transmit light of all wavelengths less than 560 nm and reflect all wavelengths greater than 560 nm. By use of a 560 nm dichroic mirror, the fluorescence from fluorescein can be split from phycoerythrin fluorescence and each one directed toward a different photodetector.

By the use of filters with precise color specifications immediately in front of the photodetectors, we have our last chance for making sure that the only

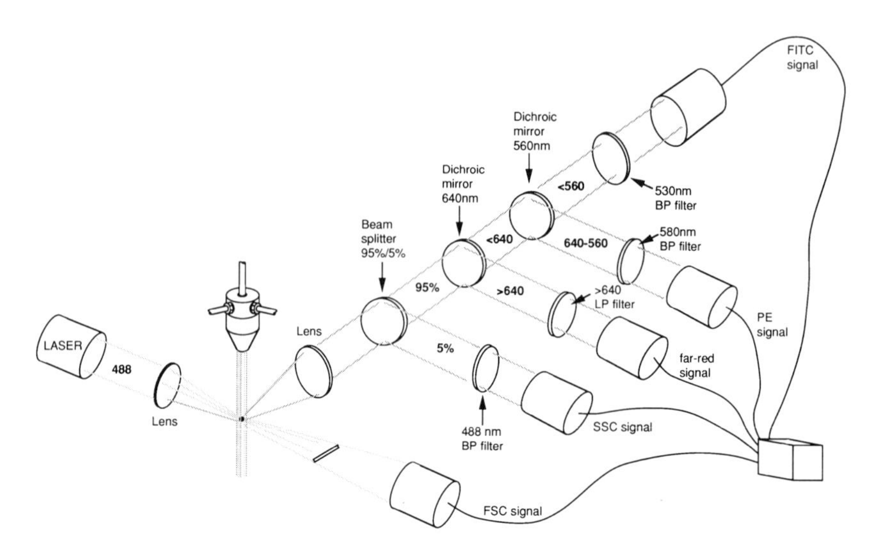


**Fig. 5.8.** A filter and mirror configuration for making five photodetectors specific for registering FSC, SSC, fluorescein, phycoerythrin, and far-red signals.

light registered on a given photodetector has been derived from the fluorescence of a given fluorochrome. Filters have more or less precise color specifications, depending on how much they cost. They come in three basic types: A short pass filter will transmit all light of a wavelength less than the specified value; a long pass filter will transmit all light of a wavelength greater than the specified value; and a band pass filter will transmit only light in a narrow band around the specified value. Band pass filters are of most use in flow systems because they are most specific in their transmission characteristics. These characteristics are usually described according to their peak transmission wavelength and the width of the wavelength band at 50% of that peak transmission. For example, if a filter transmits 90% of the 570 nm light falling onto its surface, and if this percentage of transmission drops to 45% when the wavelength changes by 15 nm in either direction, then the filter will be described as a 570/30 nm filter.

By looking at the fluorescence emission spectra of the fluorochromes in use, we can now understand the strategy that is used for selecting the filter and mirror combinations that will allow us to correlate the signal from a given photodetector with the fluorescence from a given fluorochrome.

Figure 5.9 shows the fluorescence emission spectra from fluorescein and phycoerythrin drawn on the same axes. (These two fluorochromes are the most useful example for our purposes here, but the same principles and filter strategy would apply to any group of two or more fluorochromes used simultaneously.) By inspection of these spectra, we can see that the best dichroic mirror to use for splitting FITC fluorescence from phycoerythrin fluorescence is a 560 nm mirror. This will send most of the FITC fluorescence straight through to one photodetector and reflect most of the phycoerythrin fluorescence toward a different photodetector. By fitting the FITC detector with a 525/30 nm filter and the phycoerythrin detector with a 585/40nm filter, we now have a system that will register FITC fluorescence and phycoerythrin fluorescence mainly on different photodetectors, and we would be entitled now to call one photodetector the FITC detector

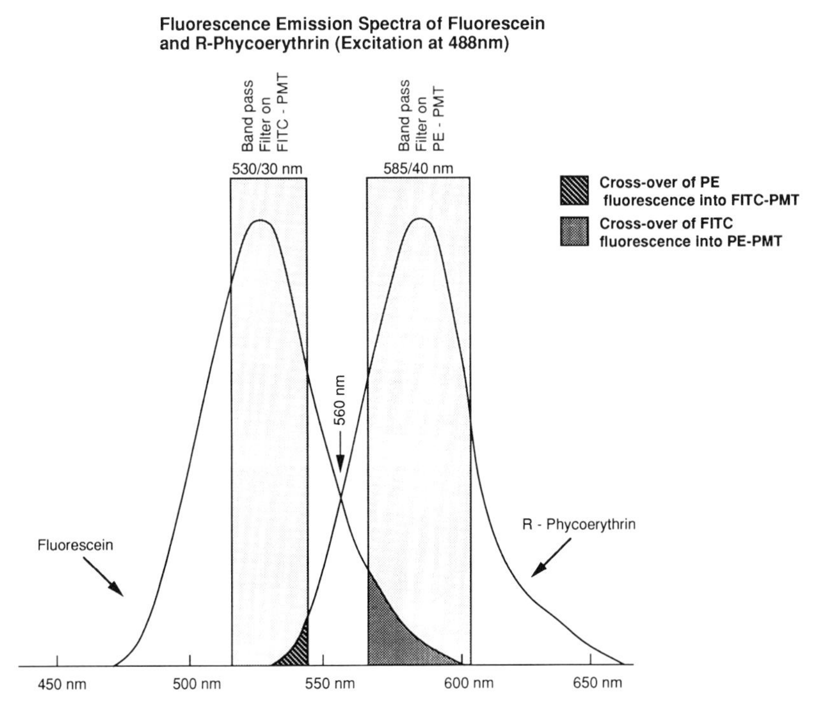


**Fig. 5.9.** The cross-over of fluorescein and phycoerythrin fluorescence through the filters on the "wrong" photodetectors.

and the other one the phycoerythrin detector. But that qualifying word *mainly* lies at the root of our next problem.

## COMPENSATION

A cell stained only with a fluorescein-conjugated stain will emit fluorescence that will register strongly on the FITC-specific photodetector (and this is fine if we are using only fluorescein as the stain in our system); however, inspection of the fluorescence emission spectrum of fluorescein (Fig. 5.9) indicates that some of the light emitted by a particle stained only with fluorescein will manage to evade our strategic obstacles (mirrors and filters); some of the light emitted by fluorescein is of a wavelength long enough to register on the phycoerythrin photodetector. No matter how tightly we restrict the filter and mirror specifications, we cannot get round the fact that there is a significant degree of overlap between the fluorescence emission of FITC and that of phycoerythrin. Therefore, there is no way to keep all the FITC light from the phycoerythrin detector without losing most of the phycoerythrin fluorescence as well. If we want to measure PE fluorescence with reasonable sensitivity, then the phycoerythrin photodetector will also respond to a limited extent to FITC fluorescence. Our use of filters and mirrors must, for these reasons, be a compromise.

What this compromise means, in practice, is that a cell stained only with fluorescein will register a signal brightly on the FITC-specific photodetector but also significantly on the phycoerythrin-specific photodetector. Similarly, a cell stained only with phycoerythrin will register a signal brightly on the phycoerythrin detector, but also slightly on the FITC detector. The left panel of Figure 5.10 shows how these pure signals look on contour plots giving the relative intensity channels for the signals on the phycoerythrin and FITC photomultiplier tubes.

Having understood why there is cross-over or overlap between fluorochromes, each registering to a certain extent on the "wrong" photodetector, we now must decide what to do about it. What we do about it is called *compensation.* Cytometers are provided with an electronic circuit, a compensation network, that measures the intensity of the signal on one photodetector and subtracts a certain percentage of that signal from the signal on the other photodetector (because a certain fixed percentage of any fluorescein signal will always cross over and register on the phycoerythrin phototube and a certain fixed percentage of any phycoerythrin signal will always cross over and register on the FITC phototube).

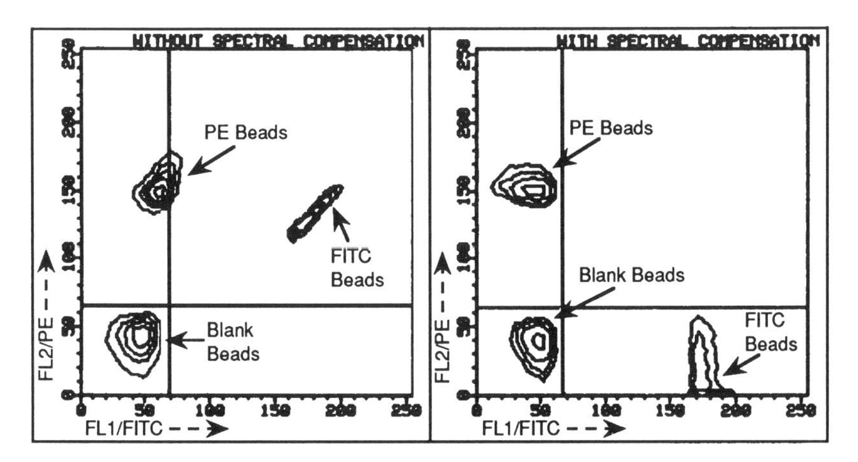


**Fig. 5.10.** The fluorescence from phycoerythrin beads registers a bit on the fluorescein photodetector; and the fluorescence from fluorescein beads registers considerably on the phycoerythrin photodetector. Compensation (right) allows us to correct for this cross-over.

Those percentages, rather than being determined spectroscopically, are routinely determined empirically. Cells or calibration beads stained with only fluorescein are run through the cytometer; their signals on both the FITC and phycoerythrin photodetectors are recorded; and the percentage subtraction (percentage of the signal on the FITC tube to be subtracted from any signal on the phycoerythrin tube) is varied until no signal above background is registered on the phycoerythrin photodetector. Similarly, we then go through the same procedure with cells or beads that have been stained only with phycoerythrin in order to determine what percentage subtraction is required to compensate the FITC phototube for cross-over from the phycoerythrin stain. And when both phototubes have been correctly compensated for cross-over from the wrong fluorochromes, our signals, plotted as two-dimensional contour plots, should look like those in the right panel of Figure 5.10.

If three fluorochromes have been used for simultaneous three-color analysis, then the signals from all three photomultiplier tubes need to be compensated individually in the same way with respect to each fluorochrome (that is, the green photomultiplier tube needs to be compensated for cross-over fluorescence from both the red and orange fluorochromes; the orange photomultiplier tube for cross-over fluorescence from the green and red fluorochromes; and the red photomultiplier tube for cross-over fluorescence from the green and orange fluorochromes). In practice, this involves using stained beads if available or else staining cells separately with each of the three fluorochromes. The separately stained cells can then each be mixed with unstained cells; the three mixtures can be run through the cytometer and examined with respect to three different dot plots (red *vs.* green; green *vs.* orange; and red *vs.* orange) to make sure that the compensation network gives square patterns (as in Fig. 5.10, right panel) for the cells in all three plots.

It remains to be said that the compensation values are valid only for a particular pair of fluorochromes with a particular set of filters and mirrors and with particular voltages applied to a particular pair of photomultiplier tubes. If any one of those elements is altered, the required compensation values will alter as well. In general, once compensation has been set using single-stained particles with a given experimental protocol (photomultiplier tube voltages, filters, and so forth), compensation values between a given pair of fluorochromes should remain constant within that protocol from day to day. However, because compensation can affect the way two-color contour plots are evaluated, correct compensation can be of critical importance in the interpretation of data from samples that are brightly stained with one fluorochrome but weakly stained with another. For this reason, single-stained controls should always be run within each experimental group in applications where this type of interpretation is required.

Through the chapters so far, a basic cytometric system has been described—we have generated an illuminating beam, followed the fluid controls that send particles through that beam, and registered the light signals emerging from those particles as specifically as possible on individual photodetectors. We have also compensated for any lack of specificity in our photodetectors that might result from cross-over between fluorochromes. Finally, we have stored the information from the light signals in a computer storage system so that they can be analyzed in relation to each other through the computer software. We are now in a position to begin to look at ways in which such a flow system might be used. In the following chapters, applications will be described to give some indication of the varied and imaginative practices of flow cytometry. My choice of specific examples is intended to illustrate points that may be of general importance and also to illustrate a range of applications that might stimulate interest in readers who are new to the field. It is not meant, nor in a short book could it hope, to be exhaustive.

## FURTHER READING

Chapter 2 in **Ormerod** is a good, gentle discussion of fluorescence technology and light detection. Chapter 4 in **Shapiro** is slightly more detailed, and Chapter 2 in **Van Dilla et al.** follows well from there.

Chapter 12 in **Melamed et al.** and Chapter 7 in **Shapiro** discuss the uses of different fluorochromes. With respect to the physical characteristics of these fluorochromes, the ***Molecular Probes Handbook*** is a valuable source of information.

The ***Melles Griot Optics Guide*** presents a good description of filter and mirror specifications and technology.

# 6

# Cells From Without: Lymphocytes and the Strategy of Gating

In this chapter, the analysis of lymphocytes, one of the more important applications of flow cytometry, is discussed. It is an important application not because lymphocytes are intrinsically more important than other types of particles, but because they constitute a class of particles that, for several reasons, are ideally suited for analysis by flow cytometry and thus make very good use of the capabilities of the technique. Lymphocytes therefore account for a large proportion of the material analyzed by flow cytometry in both hospital and research laboratories. They also serve as good examples for describing some of the general procedures of cytometric protocol, particularly the art of gating and the requirements for controls.

Lymphocytes are well suited to flow analysis for three reasons. First, they occur *in vivo* as single cells in suspension; they can therefore be analyzed without disaggregation from a tissue mass (and no spatial information is thereby lost in the course of preparation). Second, the normal suspension of lymphocytes (*i.e.*, blood) contains a mixture of particles in addition to lymphocytes that happen to give off forward (FSC) and side (SSC) light scatter signals of various intensities, thereby allowing them to be distinguished from each other by flow cytometric light scatter parameters. Third, with the monoclonal antibody technique a vast array of specific stains for probing lymphocyte surface proteins has been created; these stains have allowed the classification of anatomically identical lymphocytes into various subpopulations according to staining characteristics that are distinguishable by means of flow analysis.

I will first describe lymphocytes in a general way for those not familiar with this family of particles. I will then proceed, using lymphocytes as an example, to a description of general staining techniques for surface markers, with attention to the need for controls. At this point I will digress slightly and discuss the way cytometry results can (and cannot) be quantified. Finally, I will discuss gating strategies, in both philosophy and practice.

## LYMPHOCYTES

Whole blood consists of cells in suspension. In a "normal" milliliter ($cm^3$) of blood, there are about $5 \times 10^9$ erythrocytes (red blood cells), which are shaped like flat disks (about 8 μm in diameter) and are responsible for oxygen transport around the body. In addition, per $cm^3$ there are about $7 \times 10^6$ white cells (leukocytes) that collectively are involved in immune responses in the organism. The leukocytes appear to be heterogeneous under the microscope and can be divided according to their anatomy into several classes. There are cells that are called *monocytes,* which are of a concentration of about $0.5 \times 10^6$ per $cm^3$; these appear as round cells (about 12 μm) with horseshoe nuclei and are thought to be responsible for presenting antigens in a way that can initiate the immune response. There are also cells that appear as 10 μm circles with large round nuclei and a narrow rim of cytoplasm; they are the lymphocytes, present in a concentration of about $2 \times 10^6$ per $cm^3$. Lymphocytes are responsible for a great variety of the functions that are known as *cellular* and *antibody* (humoral) *immunity*, as well as the regulation of those functions. The third group of leukocytes occurs in a concentration of about $5 \times 10^6$ per $cm^3$. They are distinguished under the microscope by lobular nuclei and an array of cytoplasmic granules, are therefore called *polymorphs* or *granulocytes* (including basophils, eosinophils, and neutrophils), and are collectively responsible for phagocytic as well as other immune-related activities. In addition, there are small, cell-derived particles called *platelets* (about 2μm in size), in a concentration of about $3 \times 10^8$ per $cm^3$. The platelets are involved in mechanisms for blood coagulation (Fig. 6.1 and Table 6.1).

Although the blood is an easily accessible tissue to study, because red blood cells outnumber leukocytes by about a thousand to one in the peripheral circulation, the analysis of leukocytes by any technique is difficult unless the red cells can be removed. Techniques for removing red cells can be divided into two general categories: Density gradient techniques pellet red cells at the bottom of a centrifuge tube (usually taking polymorphs with them), leaving so-called peripheral blood mononuclear cells (or PBMC; that

**TABLE 6.1. Cells in Normal Human Adult Peripheral Blood**

| Cells | No. per $cm^3$ | Percent of WBC | Diameter (μm) |
|---|---|---|---|
| Platelets (thrombocytes) | $2\text{-}5 \times 10^8$ | | 2-3 |
| Erythrocytes (rbc) | $4\text{-}6 \times 10^9$ | | 7-8 |
| Leukocytes (wbc) | $5\text{-}10 \times 10^6$ | [100] | |
| Granulocytes | | | |
| Neutrophils | $3\text{-}7 \times 10^6$ | 60-70 | 10-12 |
| Eosinophils | $0.05\text{-}0.4 \times 10^6$ | 1-4 | 10-12 |
| Basophils | $0\text{-}0.05 \times 10^6$ | 0-0.5 | 8-10 |
| Lymphocytes | $1\text{-}3 \times 10^6$ | 20-30 | 6-12 |
| Monocytes | $0.1\text{-}0.6 \times 10^6$ | 2-6 | 12-15 |

Values are from Konrad Diem (ed) (1962). Documenta Geigy Scientific Tables, ed 6. England: Ben Johnson.

is, monocytes and lymphocytes) at an interface from which they can be collected. Red cell lysis techniques involve the preferential lysis of erythrocytes by the use of ammonium chloride or short exposure to distilled water, leaving intact the more robust white cells. Both techniques can be criticized for the theoretical possibility of selective (but undefined) loss of certain classes of lymphocytes, and some work has been done comparing the two methods. In general, labs seem to stick to the methods they have been using. It should be noted that density gradient methods result in a preparation of lymphocytes and monocytes, with a large proportion of the polymorphs and red cells removed. Lytic methods remove primarily the red cells, leaving polymorphs as well as mononuclear cells intact. The red cell lysis method for white cell enrichment has become the procedure of definite choice both for the handling of potentially infected blood (because the blood preparation is quick and requires less manipulation than for density gradient methods) and for the analysis of pediatric samples (because smaller volumes of blood are sufficient with this technique).

The light scatter signals (FSC and SSC) resulting from the flow cytometric analysis of whole blood, lysed whole blood, and a mononuclear cell preparation from the density gradient separation of whole blood are compared in Figure 6.2. Each dot plot shows 2,000 cells; each cell is represented by a dot and plotted on the two axes according to the intensity of its forward

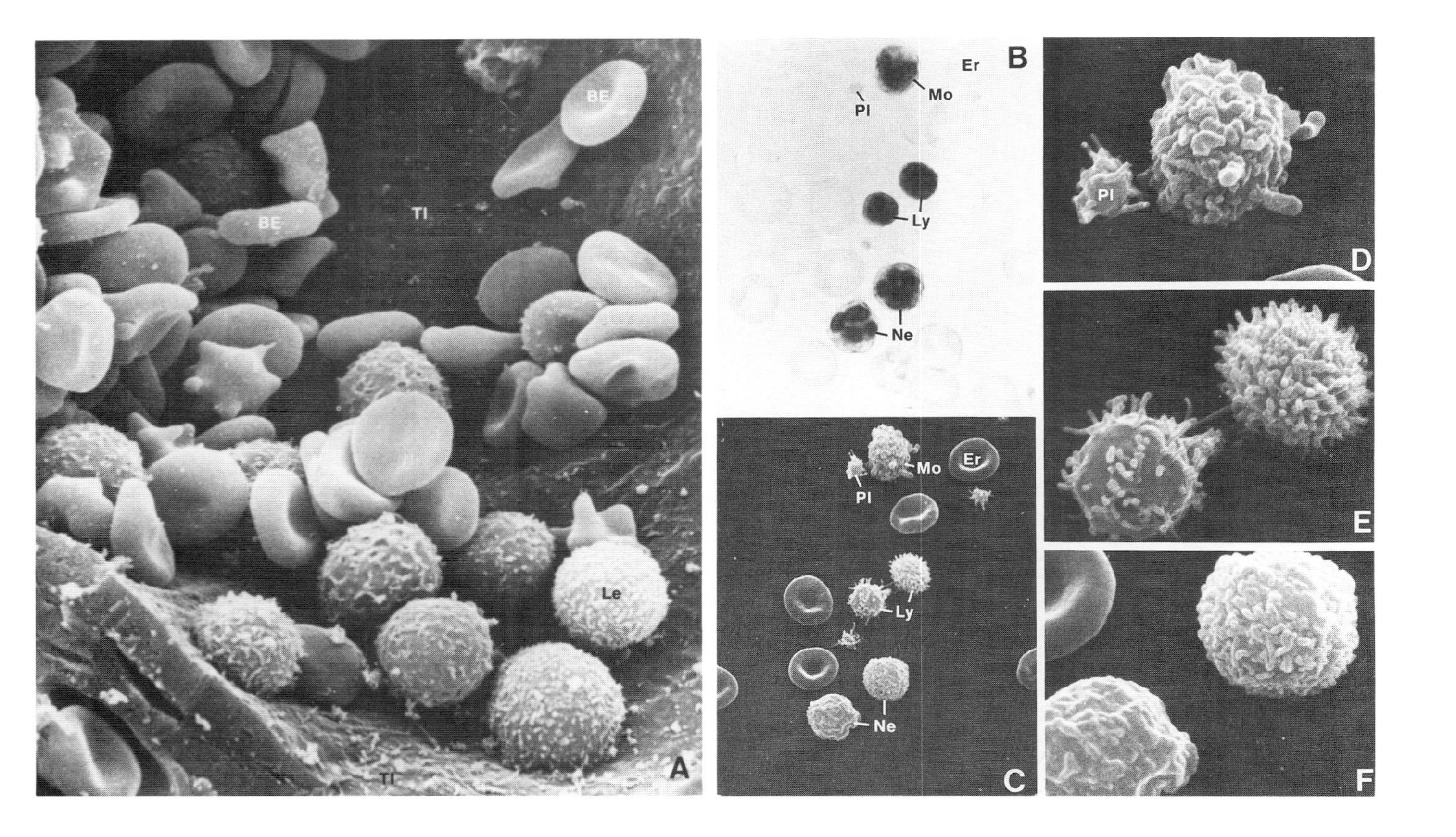
BE
BE
TI
Le
TI
A
B
Er
Mo
Pl
Ly
Ne
C
Er
Mo
Pl
Ly
Ne
D
Pl
E
F

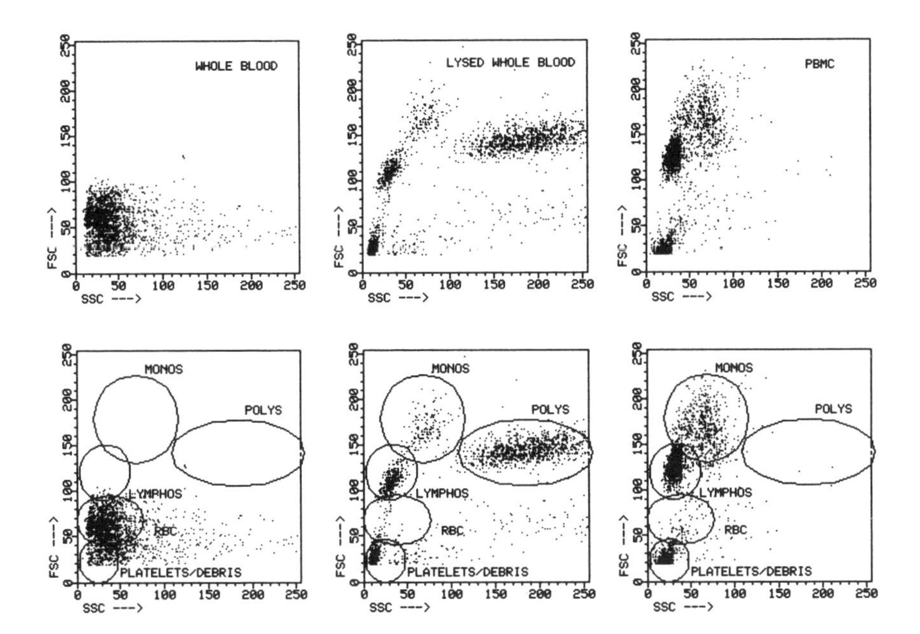


**Fig. 6.2.** The FSC and SSC signals resulting from the cells in different blood preparations. The bottom panels indicate the five clusters into which the signals fall.

scatter and side scatter signals. The resulting clusters of particles consist of 1) a group of particles with low side scatter intensity and low forward scatter intensity (and usually ignored because they fall below the useful FSC threshold); 2) a group of particles with low side scatter intensity but somewhat higher forward scatter intensity; 3) a group of particles, still with low side scatter intensity but with moderately high forward scatter intensity; 4) a group of particles with somewhat higher forward scatter, but also with moderate side scatter intensity; and 5) a group of cells possessing moderate forward scatter but much higher side scatter.

---

**Fig. 6.1.** Scanning electron micrographs showing the different surface textures of red (Er) and white (Le) blood cells. **A:** Cells within a blood vessel. **B, C:** A comparison of scanning EM with conventional light microscope images of the same field of stained cells. Enlarged pictures at the right emphasize the different surface textures of monocytes (Mo) and platelets (Pl) in **D**; lymphocytes (Ly) in **E**; and neutrophils (Ne) in **F**. From Kessel and Kardon (1979).

The most obvious way to determine the identity of these particles, with distinctive scatter characteristics, is to sort them. The flow sorting of blood preparations according to forward and side scatter parameters has brought the realization that the clusters of particles with defined flow cytometric scatter characteristics belong to the groups of cells as distinguished, traditionally, by microscopic anatomy. The anatomical differences that we can observe under the microscope are observable because they result in different patterns of light bouncing off the cells on the slide and being registered by our eyes. Therefore, it should not be entirely surprising that a flow cytometer, with its photodetectors measuring light bouncing off cells, often clusters cells into groups that are familiar to microscopists. But, in addition, the human eye registers many more than two parameters when it looks at a cell (think of area, shape, texture, movement, color). Specifically, the eye registers something we might call "pattern" very sensitively, which a flow cytometer registers not at all. So, in some ways it is perhaps just good luck that the two parameters of forward and side light scatter do allow cytometrists to distinguish some of the classes of white cells almost as effectively as a trained microscopist with a good microscope.

We should, however, always be aware of situations in which microscopists are better than cytometrists—certain types of classification are easy by eye and not at all easy by cytometer. For example, a microscopist would never confuse a dead lymphocyte with an erythrocyte, nor a chunk of debris with a viable cell, mistakes that are all too easy for a cytometrist. And microscopists, if they were not so polite, would find laughable the fact that cytometrists have a great deal of difficulty in distinguishing clumps of small cells from single large cells. But, if we think back to the discussion in Chapter 3 about the origin of the forward scatter signal, we can see why these problems occur and why we need to be aware of the limits of flow cytometry (and why a microscope is an essential piece of equipment in a flow cytometry lab).

The patterns of light scatter distribution illustrated in Figure 6.2 result from analysis of blood from a normal individual. With patterns like this, a flow cytometrist can, with some practice, set a so-called lymphocyte gate or lymphocyte box around a group of particles that are mainly lymphocytes. This lymphocyte gate will then, using forward and side scatter characteristics, define the cells that are to be selected for analysis of fluorescence staining. Such lymphocyte gating is the direct equivalent of a skilled microscopist's decision, based on size and nuclear shape, about which cells to include as lymphocytes in a count on a hemocytometer slide. A detailed discussion of gating is given at the end of this chapter; because staining can provide us with some help in our gating decisions, we need first to discuss some of the staining techniques that can help us to describe cell populations.

## STAINING FOR SURFACE MARKERS

Monoclonal antibody technology has provided flow cytometrists with a large array of antibodies that are specific for various protein antigens of the lymphocyte surface membrane. Indeed these antibodies have allowed immunologists to define taxonomic subgroups of lymphocytes that are microscopically indistinguishable from each other but whose concentrations vary in ways that are related to an individual's immune status. Some of the membrane antigens define function, some define lineage, some define developmental stage, and some define aspects of cell membrane structure whose significance is not yet understood. Table 6.2 lists some of the more common antigens that have been used for lymphocyte analysis. Although in this chapter I discuss the staining of lymphocyte surface antigens with monoclonal antibodies, the principles apply equally well to the use of antibodies for staining any surface antigens on any type of cell.

An antibody will form a strong bond to its corresponding antigen. To be of use in flow cytometry or microscopy, this bond needs to be " visualized" (to the eye or to the photodetector) by the addition of a fluorescent tag. Visualization can be accomplished by one of two different methods. With direct staining, cells are incubated with a monoclonal antibody that has been previously conjugated to a fluorochrome (most often fluorescein or phycoerythrin for the reasons of photochemistry that were discussed in Chapter 5). This procedure is quick and direct; it merely involves a single staining incubation, followed by several washes to remove weak, nonspecifically bound antibodies. Cells thus treated are then ready for flow analysis.

The second method (indirect staining) is more time-consuming, less expensive, and either more or less adaptable depending on the application. Indirect staining involves the incubation of cells with a nonfluorescent monoclonal antibody, washing to remove nonspecifically bound molecules, and then a second incubation with a fluorescent antibody (the so-called second layer) that will react with the general class of monoclonal antibody used in the first layer. For example, if the primary monoclonal antibody happens to be an antibody that was raised in a mouse hybridoma line, it will have the general characteristics of mouse immunoglobulin; the second layer antibody can then be a fluoresceinated (fluorescein-conjugated) antibody that will react with any

**TABLE 6.2. Surface Antigens Useful for Classification of Lymphocyte Subpopulations**

| Antigen designation | Predominant cell reactivity | % Positive of normal lymphocytes | Comments |
|---|---|---|---|
| CD1 | Thymocytes | <1% | Found in some T-cell malignancies |
| CD2 | T cells | 78-88% | Pan T cell marker |
| CD3 | T cells | 68-82% | Pan T cell marker |
| CD4 | Helper T cells | 35-55% | HIV receptor; low in AIDS patients |
| CD5 | T cells | 65-79% | Expressed dimly on some normal B cells; high on B CLL lymphocytes |
| CD6 | T cells | approx. 75% | Pan T cell marker |
| CD7 | T cell subset | approx. 50% | Found on almost all T ALL |
| CD8 | Cytotoxic T cells | 20-36% | CD4/CD8 ratio used to monitor immune status |
| CD10 | B cell precursors | <10% | Diagnosis of C-ALL |
| CD11 | Some T cells, NK, monos, and polys | 25-35% | |
| CD13 | Polys, monos | <10% | |
| CD14 | Monos | <10% | Useful for counting/excluding monos from gate |
| CD15 | Polys | <10% | |
| CD16 | NK | 8-22% | Useful for counting natural killer cells |
| CD18 | Leukocytes | >95% | |
| CD19 | B cells | 5-15% | Pan B; diagnosis of B cell malignancies |
| CD20 | B cells | 5-15% | Pan B; diagnosis of B cell malignancies |
| CD21 | Mature B cells | 5-15% | Typing of B cell malignancies |
| CD22 | B cells | 5-15% | Pan B; Diagnosis of B cell malignancies |
| CD25 | Activated T,B cells | 15-40% | IL2 receptor; marker for lymphoblasts |
| CD30 | Reed-Sternberg cells; Activated T and B cells | | Diagnosis of Hodgkin's lymphoma |
| CD45 | Leukocytes | >95% | Leukocyte common marker; various forms may mark memory and naive cells |
| CD56 | NK cells | 9-21% | |
| HLA-DR | B cells, monos, activated T cells | 7-15% | Marker for activated T cells |

Data for normal ranges are from Becton Dickinson Immunocytometry Systems and from *Leukocyte Typing IV* (Oxford University Press).

mouse immunoglobulin. The advantages of this two-layer technique are, first, monoclonal (primary layer) antibodies are much cheaper if they are unconjugated; second, a given second layer reagent can be used to visualize any monoclonal antibody of a given class (*e.g.*, any mouse immunoglobulin, in our example here); and third, each step in the staining procedure results in amplification of the fluorescence intensity of the staining reaction. (With regard to this signal amplification that occurs with indirect staining, I might also add parenthetically that further amplification of weak signals is possible by using third and fourth layer reagents of appropriate specificity. All that is required is an interest in zoology. For example, if the primary monoclonal antibody is a mouse monoclonal, the second layer reagent could be a fluorescein-conjugated antibody raised in a goat and specific for mouse Ig [known as a *goat antimouse* reagent]. An appropriate third layer reagent might then be a fluorescein-conjugated antibody raised in a sheep and specific for goat Ig [sheep antigoat], and so on [elephant antisheep, armadillo antielephant] until the zoologists run out of immunologically competent and willing animals [Fig. 6.3]. Because each antibody molecule is linked to many fluorescein molecules and because many antibodies will bind to each antigen, this has been proposed as a way to increase the intensity of signals from very sparsely expressed membrane proteins.)

The disadvantages of indirect staining are that it is more time-consuming and it involves a second step that doubles the opportunity for nonspecific binding. Indirect staining also greatly limits the opportunity for simultaneous double and triple staining of cells with two and three different fluorochromes because of the problems of cross-reactivity between primary antigens and the conjugated second layer reagents that may display broad specificity. Nevertheless, with appropriate choice of monoclonal antibodies of specific immunoglobulin subclass and/or animal derivation and with second layer reagents appropriate and specific to these particular characteristics, double staining with indirect reagents is sometimes possible—but it is not easy.

## CONTROLS

One of the general laws of science that particularly applies to flow cytometry is that no matter how many controls you have used in an experiment, when you come to analyze your results you always wish you had used one more. There are three reasons that this problem is often encountered in flow research. One has to do with the background fluores-

Fig. 6.3. Amplification of staining by the use of multiple antibody reagents. Drawn by Ian Brotherick.

cence of unstained cells; the second has to do with the nature of antibody-antigen interactions; and the third has to do with the problem of compensation between overlapping fluorochrome spectra.

The first problem that needs to be controlled is that of background fluorescence (called *autofluorescence*). All unstained particles give off some fluorescence (that is, all particles will emit some light that gets through the filters in front of a cytometer's photodetectors). This autofluorescence may not be recognized by microscopists, either because it is very dim or because experienced microscopists have acquired a mental threshold in the course of their training. But our cytometer's photodetectors are both very sensitive and completely untrained. Therefore the autofluorescence of cells, resulting from intracellular constituents such as flavins and pyridine nucleotides, is bright enough to be detected (and can, in some cells, be so bright as to limit our ability to detect positive staining over and above this bright background).

Beyond the problem of autofluorescence, there is a second problem. As discussed above, much of the staining of cells for flow analysis makes use of antibody-antigen specificity. While the specificity between an antibody's binding site and the corresponding epitope on an antigen is indeed exquisite, the beauty of the system can be confounded by long floppy arms on the nonspecific (FC) end of the antibody that stick with wild abandon to so-called FC receptors that occur on the surface of many types of cells (notoriously monocytes). What this means is that cells may stain with a particular monoclonal antibody because they possess a particular antigen on their surface membrane that locks neatly with the key on the monoclonal antibody binding site. But they may also stain with that particular monoclonal antibody because they possess "sticky" surfaces that promiscuously cling to any and all antibodies. In addition, it is often true that dead cells (with perforated outer membranes) can soak up antibodies and then hang on to them tenaciously (there is more about ways to avoid the dead cell problem in Chapter 7). The way to know if staining of cells is specific to a specific antigen is to use the correct control.

The correct control is always an antibody of exactly the same properties as the monoclonal antibody used in the experiment, but with an irrelevant specificity. If we are staining human mononuclear cells with a monoclonal antibody having a specificity for the CD3 protein occurring on the surface of T lymphocytes (and that monoclonal antibody happens to be a mouse immunoglobulin of the $IgG_{2a}$ subclass, conjugated with six fluorescein molecules per molecule of protein and used to stain the cells at a concentration of 1 μg per ml), then an appropriate control would be a mouse monoclonal antibody of the same subclass, with the same fluorescein conjugation ratio, and at the same protein concentration, but with a specificity for something like keyhole limpet hemocyanin or anything else that is unlikely to be found on a human blood cell (Fig. 6.4). Such a control antibody is known as an *isotype control* because it is of the same immunoglobulin isotype (subclass) as the staining antibody used in the experiment; it will allow the scientist to determine how much background stain is due to irrelevant stickiness (dead cells, FC receptors, and so forth). The only trouble with this scenario is that the exactly correct isotype controls are not usually available. Various manufacturers of monoclonal antibodies will sell so-called isotype controls and will certainly recommend that they be used. But these are general purpose isotype controls that will be of an

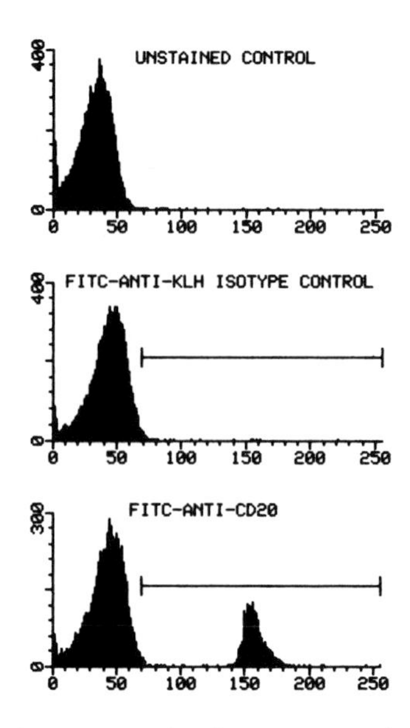


**Fig. 6.4.** The fluorescence histogram of an isotype control sample is used to decide on the fluorescence intensity that indicates positive staining.

average fluorochrome conjugation ratio and of a protein concentration that may or may not be similar to that used for most staining procedures. Whether an average isotype control is better than no isotype control at all is a matter of opinion. It certainly deserves a bit of thought.

The third problem that needs to be controlled is that of spectral cross-over and the possibility of incorrect instrument compensation. As an example of a case in which controls for nonspecific staining, autofluorescence, and compensation are all critical, let us look at the situation that pertains with the staining of B lymphocytes for the CD5 marker (remember that example from Chapter 4) present with only low density on their surface. As well as the problems created by nonspecific staining and by autofluorescence, the problem of spectral cross-over between fluorescein and phycoerythrin (PE) can particularly confuse the interpretation of results from this kind of experiment. Look at Figure 6.5. What we are interested in is the number of B lymphocytes that possess the CD5 surface antigen. These cells will appear in quadrant 2 of a contour plot of FITC fluorescence (a B-cell stain) on the horizontal axis against PE fluorescence (the CD5 stain) on the vertical axis. But B cells will also appear in this quadrant if they have orange autofluorescence or if they are nonspecifically sticky for the anti-

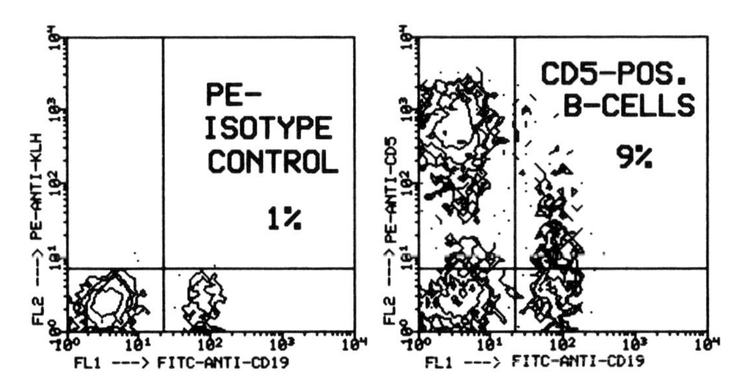


**Fig. 6.5.** The use of a PE isotype control to help in deciding where, in a dual-color plot, to draw the horizontal line between FITC-stained cells to be considered positive and those to be considered negative for the PE stain. Misplacing of the horizontal line will affect the number of CD19 cells determined to express the CD5 antigen in the stained sample. Data courtesy of Jane Calvert.

CD5 antibody (in this case a mouse monoclonal immunoglobulin of the $IgG_{2a}$ isotype). In addition, they will appear in this quadrant if the cytometer's PE photodetector has not been properly compensated for cross-over from the fluorescein signal. The way around all these problems is to stain cells with an FITC stain for B cells in conjunction with an isotype control (a mouse $IgG_{2a}$ antibody conjugated with PE but specific for an irrelevant antigen, say, keyhole limpet hemocyanin). The intensity of stain shown by these control cells on the PE photodetector will mark the limit of intensity expected from all nonspecific causes. Any further PE intensity shown by cells stained with the B-cell stain and the anti-CD5 PE stain will now clearly be the result of specific CD5 proteins on the cell surface. In this way, by use of the correct isotype control, we can rule out any problems in interpretation that may result from incorrect instrument compensation or nonspecific or background fluorescence.

In general, all these problems and their appropriate controls are particularly important when, as with the CD5 antigen on B cells, the staining density on the cells in question is low and there is considerable overlap between positive and negative populations. They become less critical for the evaluation of results when dull negative cells are being compared with a bright positive population. In any case, the general procedure for analyzing flow data is to look at the level of background staining (resulting from both autofluorescence and from nonspecific staining) and then, having defined this intensity, to analyze the change in pattern that occurs after the cells have been stained. As discussed in Chapter 4, this change may consist of the bright staining of a small subpopulation within the total population; in this situation, the relevant result may be given as the percentage of the total number of cells that are positively stained. Alternatively, the change may involve the shift of the entire population to a somewhat brighter fluorescence intensity; here the relevant result may be expressed as the change in brightness (mode, mean, or median of the distribution). This leads us to the problem of quantification of brightness by flow cytometry.

## QUANTIFICATION

One of the proclaimed advantages of flow cytometry, compared with eyeball microscopy, is its quantitative nature. Flow cytometry is indeed very quantitative when it comes to counting cells. Users are, however, subjected to a rude shock when they first attempt to quantify the fluorescence intensity of their flow data. Whereas a flow cytometer can be very quantitative about *comparing* the fluorescence intensity of particles (assuming that the photodetectors and amplifiers are working well and that log amplifiers are set correctly), it is unfortunately true that a flow cytometer is very bad at providing an absolute value for the light intensity it measures. Therefore, any experimental protocol that needs to measure the intensity of staining of cells (as opposed to a yes or no answer about whether and what percentage of cells are stained or not) is up against certain intrinsic difficulties. If you really do need some measure of the intensity of cells, the way around these difficulties is to accept the limitations of the system, work within the constraints, and use some kind of standard to calibrate the intensity scale. The easiest standard for any cell is its own unstained (or nonspecifically stained) control. An arbitrary value can be assigned to the fluorescence of the control, and the stained sample can be compared with this. The resulting relative fluorescence could be expressed as the relative modal or mean or median fluorescence of the population (modes, means, and medians will give the same value if the control and stained populations are normally distributed; each has its own advantages if the populations are not). The disadvantage in this measure of relative fluorescence compared with the control is that cells with high autofluorescence will require a greater density of positively stained receptors to give the same "relative intensity" value as cells with low autofluorescence. In other words, if intensity is expressed by a ratio of the brightness of stained cells relative to that of the unstained cells, a given ratio will represent more positive stain (in terms of fluorochrome molecules) on highly autofluorescent cells than on cells with low background. Another way of saying this is simply to note that our ability to detect stain is more sensitive on cells with low background levels (and this, of course, corresponds to our intuitive feel for the subject).

One way around this problem is by comparing cell fluorescence with the fluorescence of an external standard. This can be done by the use of calibrated fluorescent beads. In brief, there are commercially available polystyrene beads (or microspheres) that have specified numbers of fluorochrome molecules bound to their surface (fluorescein beads are most easily prepared and most stable, but beads with other fluorochromes are also available). By running the calibrated beads through the cytometer, a calibration curve can be obtained (Fig. 6.6), giving each channel on the ADC an equivalence in number of fluorochrome molecules (or, more exactly, in equivalent soluble fluorescein molecules). In this way, the background fluorescence of a control sample (in terms of equivalence as number of soluble fluorochrome molecules) can be subtracted from the number of fluorochrome molecule equivalents of a stained sample and the fluorescence of the stained sample expressed as, for example, fluorescein molecules over and above the background level.

Having now determined a value that will quantify the brightness of a particle in terms of fluorochrome equivalents, one may wonder how best to convert that figure into the number of receptors or antigens on the surface of the cell. Calculation of this value might be easily determined if values are known for the number of fluorochrome molecules per antibody (the F/P ratio) used in the staining procedure. Unfortunately, even if this value has been determined chemically, it may not apply within a staining system in which there is masking or shading in closely packed regions on a cell surface and in which the conjugated fluorochrome may fluoresce less brightly than its soluble form. Moreover, the F/P value will almost certainly not be known in a system with indirect staining and undetermined amplification. However, help may be available in the form of a different type of calibrated microsphere. Beads can be obtained that possess a known number of binding sites for immunoglobulin molecules. They can be treated as if they were cells and stained in the routine way with a stain (direct or indirect) in question. The number of fluorescein molecules binding to the specified number of sites on the bead surface can then be determined by reference to the calibration curve derived from the fluorochrome-conju-

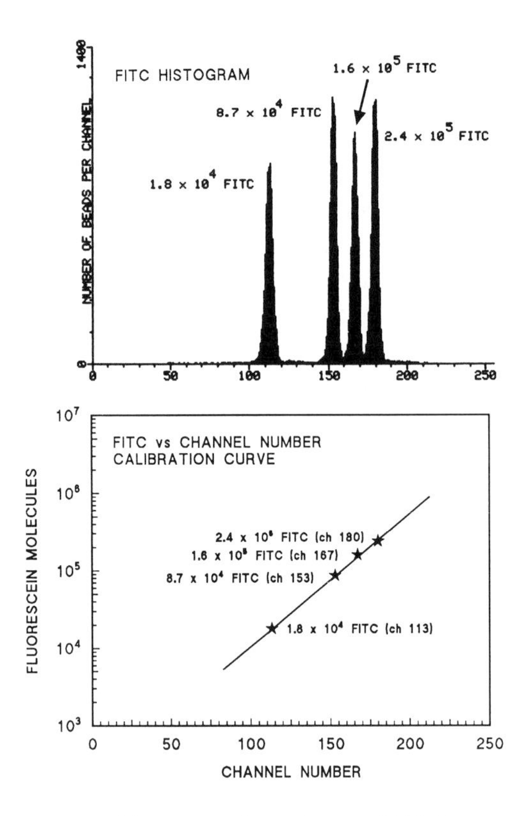


**Fig. 6.6.** The fluorescence histogram of a mixture of FITC-conjugated calibration beads and the calibration line for channel numbers and their equivalence in soluble fluorescein molecules derived from that histogram.

gated beads. This value will thus allow the calculation of the effective fluorochrome-to-binding-site ratio for that particular stain on the surface of a cell.

## THE STRATEGY OF GATING

In flow cytometry the term *gating* is applied to the selection of cells (according to their fluorescence and/or scatter characteristics) that will be carried forward for further analysis. The process of gating corresponds to the decisions made by the microscopist about what particles in a field to include in a count of any particular type of cell. Gating is generally acknowledged to be one of the most problematic aspects of flow cytometry. It is problematic, however, primarily because flow cytometrists like to think of their technique as objective and do not like to admit that much of flow cytometric analysis rests upon the foundation of a few subjective decisions. The ideal strategy for gating should therefore move us toward two goals: First, gating needs to become as objective as possible, and, second, flow cytometrists need to recognize explicitly those aspects of gating that continue to require subjective decisions.

To continue with the example drawn from lymphocyte techniques, gating has been employed effectively in the use of FSC and SSC characteristics for the selection of lymphocytes from within a mixed population of leukocytes from peripheral blood. This use of gating is derived from two causes: First, the usual questions asked by immunologists (whether of a microscope or of a flow cytometer) concern the distribution of lymphocytes into subpopulations. For example, "what percentage of lymphocytes are B lymphocytes?" or "what percentage of lymphocytes are CD4-positive lymphocytes?" Second, as discussed at the beginning of this chapter, lymphocytes possess physical characteristics that generally allow them to be distinguished from other types of leukocytes in both flow and microscopic analysis. Therefore, if one wants to know what percentage of the lymphocytes are B lymphocytes, one simply defines the FSC and SSC characteristics of lymphocytes (by flow) or the nuclear and cytoplasmic patterns of lymphocytes (by microscope) and then analyzes the particles within this gate (*i.e.,* with the defined characteristics) to see which of these stain with an anti-CD19 monoclonal antibody more intensely than an unstained control (Fig. 6.7). The ability to define a lymphocyte gate takes a bit of practice, but it is usually an easy decision for a skilled flow cytometrist (and also for a skilled cytologist using a microscope) when blood is relatively normal. It can, however, be a very difficult decision, by either technique, when blood is abnormal (for example, in blood from immunosuppressed patients, who have relatively few and possibly activated lymphocytes).

Figure 6.8 shows the light scatter profiles from mononuclear cell preparations from patients with immunosuppression and from patients with a lymphocyte leukemia, just to show two extreme examples of possible light scatter profiles. Figure 6.9 also indicates the change in light scatter that occurs when lymphocytes become activated or stimulated to divide; both forward and side scatter intensity increase, and the lymphocytes "move out of the lymphocyte box." When few lymphocytes are present or those that are present have been enlarged through immunological challenge, deci-

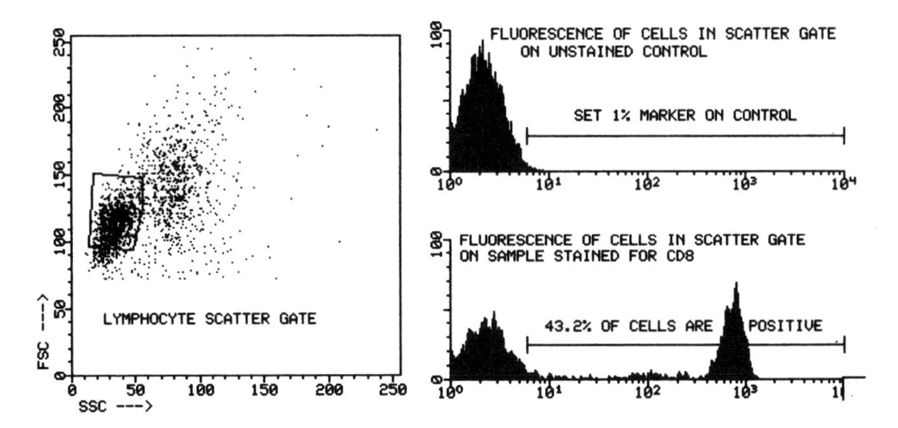


**Fig. 6.7.** A common type of flow analysis. The cells within a lymphocyte gate are analyzed, first in an unstained control sample and next in a stained sample, to determine how many cells within that gate are positively stained.

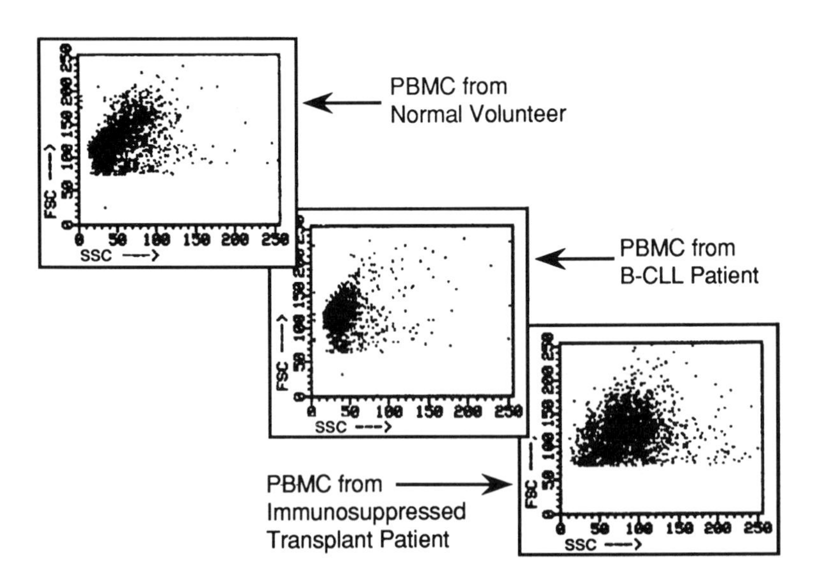


**Fig. 6.8.** Peripheral blood mononuclear cells from a normal volunteer, a B CLL leukemic patient, and a kidney transplant patient. Normal PBMC appear as a tight cluster of cells (lymphocytes) with a diffuse group of monocytes on their shoulder possessing brighter FSC and SSC. The leukemic patient's cells are almost exclusively lymphocytes. PBMC from an immunosuppressed transplant patient contain few lymphocytes; the placement of a lymphocyte scatter gate can be difficult.

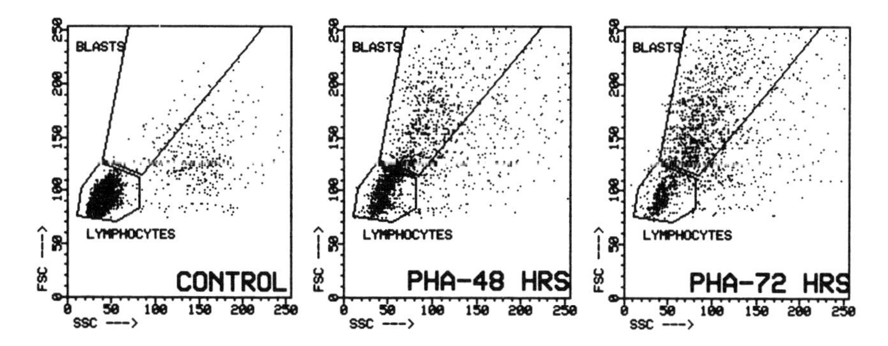


**Fig. 6.9.** The changes in patterns of scattered light that occur when lymphocytes become activated.

sions about what is or is not a lymphocyte become difficult, both for microscopists and for cytometrists.

The most important message here is that this type of decision, whether it be microscopic classification or a cytometrist's gating, is essentially a subjective decision. It is based on training, skill, and practice. Its subjective nature can lead to different results from different operators, particularly when operators are inexperienced or when abnormal specimens are being analyzed, because the percentage of cells within a gate that are stained (with any given stain) will vary depending on the range of cells included within that gate (see Table 6.3). If we want to analyze lymphocytes, then the gate should ideally include all the lymphocytes from the sample and should include only lymphocytes.

Recognition of the need for some objectivity in gating decisions (both because operators may not be experienced and because not all samples are "normal") has led to attempts both to automate the defining of the gate and to describe the goodness of the gate, once defined. Progress toward automation with lymphocytes brings us to the technique of back-gating, which has proven valuable as a strategy for flow cytometric analysis in general. Historically, gating has been performed on the SSC and FSC characteristics of cells (by analogy with a microscopist's use of physical criteria to classify cells), and this gate has then been used to enquire about the fluorescence properties of the gated cells. Back-gating is really just a different sort of gating—it reverses the usual protocol by using a stained sample, placing a gate around particles with certain fluorescence characteristics, and then

**TABLE 6.3. The Effect of Gate Size on Determination of Characteristics of Cells Within That Gate: Peripheral Blood Mononuclear Cells From a Heart/Lung Transplant Patient**

| | Small gate (691 cells) | Large gate (1074 cells) |
|---|---|---|
| Lymphocytes | 646 cells<br>94% of gated cells | 928 cells<br>86% of gated cells |
| CD3 cells (T cells) | 81% of gated cells | 70% of gated cells |
| IL-2r + (activated) T cells | 12% of T cells | 16% of T cells |
| CD20 cells (B cells) | 9% of gated cells | 8% of gated cells |
| CD4 cells (helper T cells) | 56% of gated cells | 51% of gated cells |
| CD8 cells (cytotoxic T cells) | 24% of gated cells | 20% of gated cells |
| CD16 (NK cells) | 8% of gated cells | 11% of gated cells |

asking what the physical characteristics (*i.e.*, FSC and SSC) of these particles are. Once the scatter characteristics of a population have been ascertained in this way, the subsequent placing of the scatter gate can be done on the basis of rational and defined criteria. For example, if one stains leukocytes with a PE-conjugated monoclonal antibody specific for the CD14 determinant (a monocyte marker) and puts a gate around the cells that are brightly fluorescent, it can be determined where these bright cells fall in a plot of FSC vs. SSC (Fig. 6.10). As it turns out, most of the $CD14^+$ cells have moderate SSC and moderate FSC and lie in a cluster just about where we have identified monocytes to be (by sorting the FSC/SSC clusters). But it also turns out that there are a few $CD14^+$ cells that lie outside this region. We are now in a position to count the total number of $CD14^+$ cells in the sample; to draw what we think is an appropriate gate within the FSC/SSC plot; and finally to ask two questions that will tell us how good a gate we have drawn: 1) how many of the $CD14^+$ cells have we excluded by the drawing of that gate? and 2) what percentage of the cells within the gate are not monocytes? So the technique of back-gating has allowed us to make an "educated guess" in drawing a monocyte gate in the FSC/SSC plot; and it has then allowed us to evaluate that gate in terms of both purity and inclusivity. A perfect monocyte gate would contain only monocytes (100% purity) and would include all the monocytes (100% inclusivity). As we shall see, most gates are a compromise between these two goals.

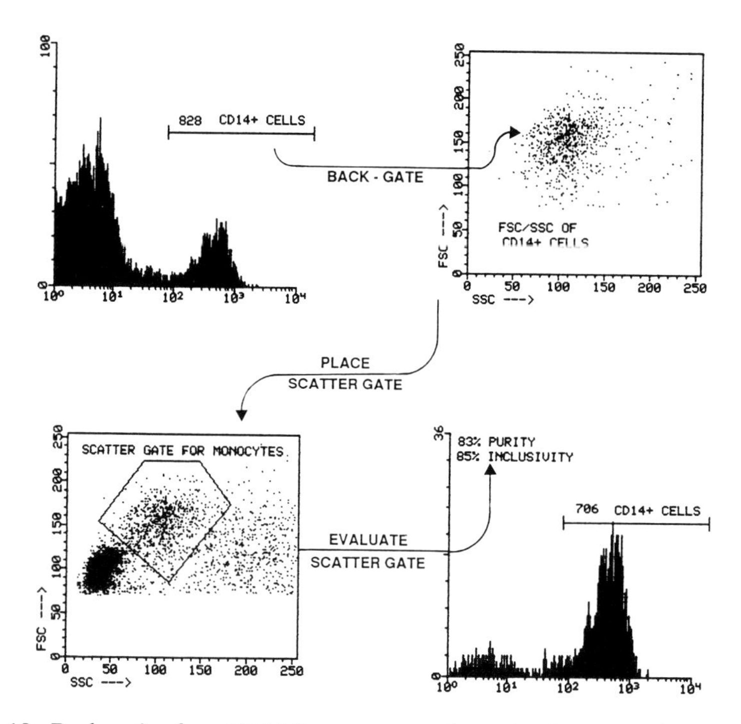


**Fig. 6.10.** Back-gating from CD14 fluorescence to determine the scatter characteristics from monocytes. Such back-gating facilitates the placing and then evaluation of a monocyte scatter gate.

By extension from this simple example, the technique of back-gating has been applied quite elegantly to leukocytes by the use of a combination of two stains and two-color analysis. When a PE-conjugated monoclonal antibody specific for monocytes (as described above) is mixed with a fluorescein-conjugated antibody specific for a determinant found on all white cells (CD45 is a so-called leukocyte common antigen), this mixture can be used to stain a white cell preparation. After dividing the FITC/PE plot into four quadrants, it can be imagined that any erythrocytes or debris in the preparation will appear in the lower left quadrant (double negatives); any white cells will appear in the upper or lower right quadrants (FITC positive); and any monocytes will appear in the upper right quadrant (double positive). One might imagine that lymphocytes and granulocytes would appear in the lower right quadrant, as they are leukocytes but not monocytes (and should be FITC positive, PE negative).

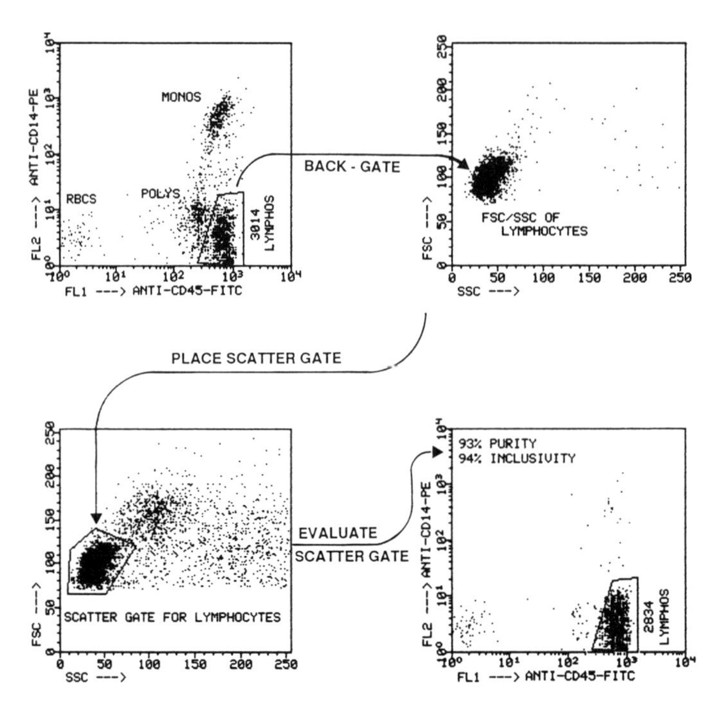


**Fig. 6.11.** Back-gating from CD14/CD45 fluorescence to determine the scatter characteristics of lymphocytes. Such back-gating facilitates the placing and then evaluation of a lymphocyte scatter gate.

The results from such a staining protocol are given in Figure 6.11. It turns out that they are even more useful than we might have imagined: It happens that granulocytes express a lower density of the common leukocyte antigen on their surface than do lymphocytes, and these two types of cells can be distinquished from each other as well as from the PE-positive monocytes in this staining mixture. Thus this protocol allows us to do the two things we have set as goals for our lymphocyte gating strategy. We can back-gate from the lymphocyte cluster in the FITC/PE plot to help us find the FSC/SSC region in which to draw our gate, and we can use the staining profile to evaluate that gate in terms of purity and inclusivity. In the case of the example shown in Figure 6.11, back-gating from the FITC-bright/PE-negative cluster leads us to a region for the FSC/SSC gating of lymphocytes. If we draw a reasonable scatter gate based on the scatter signals of the lymphocytes, we find that we have included most of the lymphocytes and that almost everything in that gate is a lymphocyte. A smaller gate might have higher purity but miss out some lymphocytes; a larger gate might include 100% of the lymphocytes but some monocytes, polys, and red blood cells as well. This kind of staining and back-gating protocol has led the way toward a fully automated gating procedure, in which the gating and evaluation of that gate are done by computer. A computational gating algorithm can be written to adjust the scatter gate to aim for some desired level of purity and/or inclusivity.

However, any automated procedure will have some degree of difficulty with abnormal samples, and laboratories will tend to have their own (subjective) methods for compromising when the sample is such that a gate with both high purity and high inclusivity is not possible. By way of illustration of this situation, we can look at a sample of blood cells from an immunosuppressed patient (Fig. 6.12). Back-gating from the FITC-bright/PE-negative cluster leads us to placing a gate in a region of the FSC/SSC plot. But we can see that there appear to be lots of lymphocytes that have abnormally high FSC and SSC; these could be activated lymphocytes. If we draw a gate large enough to include these lymphocytes in our subsequent analysis (high inclusivity), we find that the gate also includes many monocytes (low purity).

Unfortunately there is really no entirely satisfactory way out of this dilemma—the simple fact is that activated lymphocytes look rather similar to monocytes. Any automated software will have trouble handling this type of situation, because it is impossible to draw a gate with both high purity and high inclusivity. Practice will differ from lab to lab, with some operators tending to aim for high purity and accepting low inclusivity while others aim in the opposite direction. In either case, if one is expressing results as the percentage of lymphocytes that stain with a given marker, these results should be corrected for contamination by other particles (*e.g.*, monocytes) within the lymphocyte gate. In fact the situation is even more complicated than that. Because one may be staining lymphocytes for markers that appear on, for example, only activated cells (*e.g.*, the IL-2 receptor), the inclusion or exclusion of the larger, more activated cells in the lymphocyte gate may have a profound effect on the result obtained even after this correction. Thus, the upshot is that we can aim for objectivity, but our decisions are often, by necessity, a subjective compromise between conflicting goals. Back-gating at least allows us to define explicitly the terms of this compromise (by quantifying the purity and inclusivity of a gate).

With the knowledge of these methods for placing FSC/SSC gates and for defining their goodness (and also having seen why gating is often

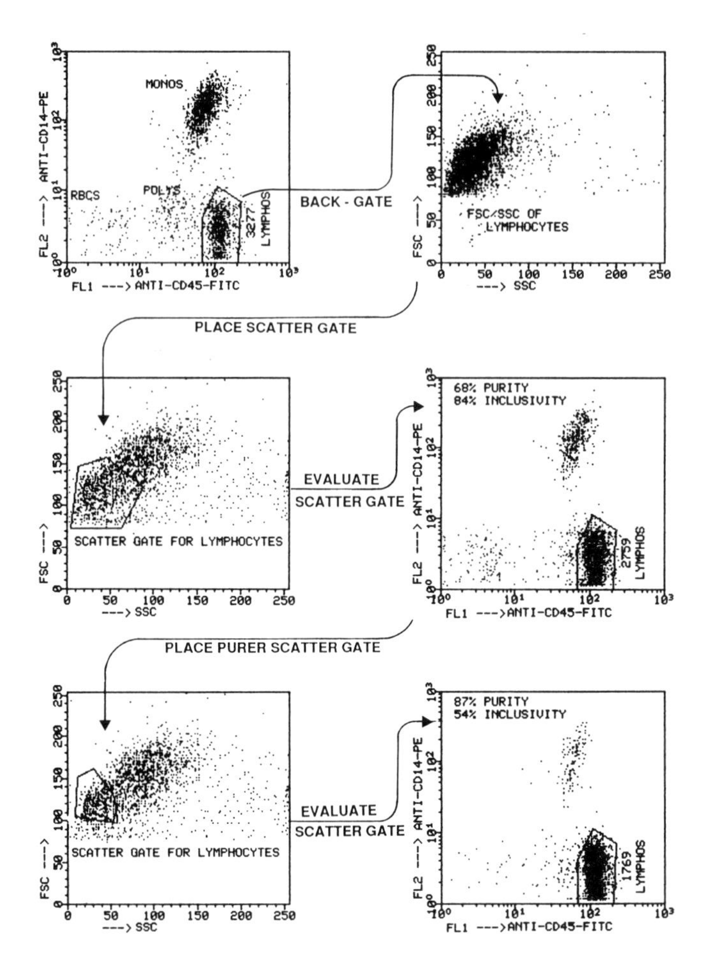


**Fig. 6.12.** Difficulties in placing a pure and inclusive lymphocyte scatter gate on blood from a transplant patient with few and blasted lymphocytes.

difficult), we are now in a position to see why current practice leads us away from scatter gating toward a quite different strategy for flow analysis. The basic problem, as seen above, is that lymphocytes (or any other taxonomic group) are not a homogeneous collection of cells with perfectly delineated physical characteristics; they are mainly homogeneous, but they usually contain at least some cells at the fringes with unusual appearance. Any gating based on FSC/SSC forces us to exclude these fringe cells. With the availability of multicolor instrumentation and multicolor stains, it has become possible to avoid making any of these difficult gating decisions and to include all cells in the analysis by using stain itself either to gate in or gate out the cells of interest. For example, if one is studying the prevalence of a particular subpopulation among lymphocytes, one could stain every PBMC sample with a PE stain for monocytes in addition to an FITC conjugate of the marker of interest. Then, in the analysis stage, one could simply gate out (exclude) from analysis any PE-positive particle and analyze all the PE-negative particles for the percentage that are FITC positive. Figure 6.13 shows an example of this protocol. By the use of three-color analysis, there are even greater possibilities. This type of analysis avoids the necessity of prior decisions about the FSC and SSC characteristics of the cells of choice (*e.g.*, lymphocytes). It becomes even more important when the cells of interest are less homogeneous in physical characteristics than lymphocytes: for example, by gating on a stain that is specific for cytokeratin (a protein found on tumor cells of epithelial origin), tumor cells within a mixed population from a breast tumor biopsy specimen can be selected for further analysis (*e.g.*, DNA content) without any prejudgement about the FSC or SSC of a poorly defined and heterogeneous population of abnormal cells.

A logical extension of this kind of technique can be seen in the methodology proposed by Horan: No decisions based on the scatter characteristics of cells have been made. Cells are stained simply with a cocktail of conjugated monoclonal antibodies at appropriate concentration, and all the

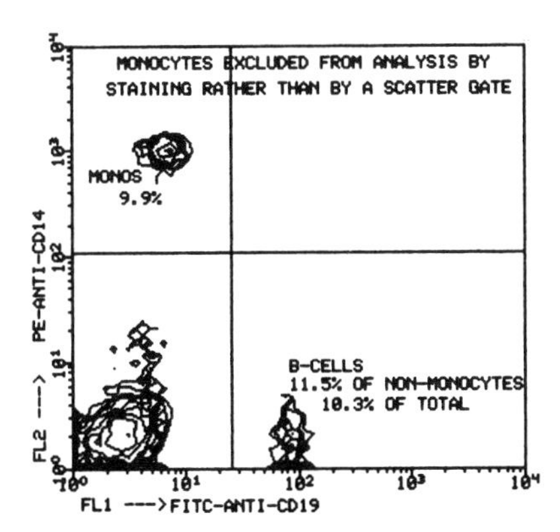


**Fig. 6.13.** Rather than a scatter gate, PE-anti-CD14 stain can be used to exclude monocytes from a count of B and non-B lymphocytes. Data courtesy of Jane Calvert.

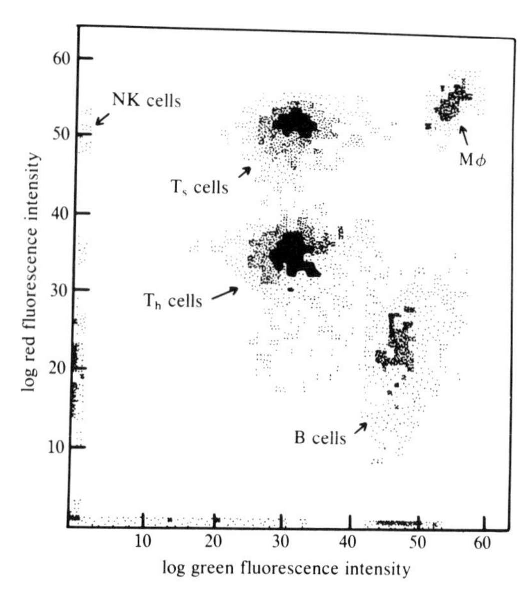


**Fig. 6.14.** The two-color fluorescence profile of PBMC stained simultaneously with six different monoclonal antibodies to delineate five different populations of cells. From Horan et al. (1986).

cells in the mixture are classified according to their staining characteristics (Fig. 6.14).

At the beginning of this section, a strategy was described for placing and then evaluating a scatter (FSC/SSC) gate (in terms of purity and inclusivity). We were then forced to admit that gating is often an uneasy and subjective compromise between these conflicting criteria. We therefore find that our argument has taken us to the conclusion that using FSC and SSC characteristics to gate cells may not be a good thing after all. The availability of multicolor analysis has led to a trend toward using staining characteristics to define the cells of interest without regard to their possibly variable physical scatter properties. By using one or more colors in the analysis either to select or to exclude (gate in or gate out) particular groups of cells, we can avoid any prejudgement on the physical characteristics of those cells. This overall strategy makes sense when our system allows for multicolor analysis and when markers are available to define the cells of interest.

As an overall summary comment on gating, we need simply to remember that in flow cytometry our questions are usually formulated in terms of "what percentage of a certain population of cells is positive for a certain set of characteristics?" The choice of a gate defines that "certain population" of cells; and the choice of that gate will therefore affect the answer to the question (the percentage positive). Whether gating is applied by means of scatter characteristics, by means of staining characteristics, or not at all, the procedure still needs to be described and quantified if the results are to be meaningful and reproducible. It is only when we have stated exactly which "certain population of cells" we are analyzing that we have fulfilled our goals of objectivity and/or *explicit* subjectivity in flow analysis.

## FURTHER READING

Chapter 3 in **Ormerod** and Chapters 17 and 34 in **Melamed et al.** are good discussions of general lymphocyte staining methodology for flow analysis. Volume I of **Weir** has several articles on the application of fluorescence immunochemistry to lymphocyte analysis.

The ***Purdue Handbook*** is a useful collection of practical protocols, as is the book by **Darzynkiewicz and Crissman.**

The ***Leukocyte Typing IV*** report is a thorough compendium of the characteristics of the leukocyte surface antigens as defined by monoclonal antibodies.

# 7

# Cells From Within: DNA and Molecular Biology

In the previous chapter I discussed how it is possible to stain protein markers on the surface of cells and then to analyze these cells for the presence and intensity of that stain by the use of flow cytometry. In addition to protein, the other biochemical component that can be used to classify different types of cells is, of course, DNA: It should therefore come as no surprise that flow cytometrists have developed methods for analyzing the DNA content of cells.

## FLUOROCHROMES FOR DNA ANALYSIS

Several types of fluorescent stain are available for the analysis of DNA; their characteristics make them suitable for different applications (see Table 7.1). The most specific stains (*e.g.*, Hoechst 33258, which stains specifically for AT groups on DNA) require the use of a laser with significant UV output. Hoechst 33258 also penetrates the outer membrane of living cells and can therefore be used for staining and then sorting living cells with different DNA content for subsequent separate culture or functional analysis. Chromomycin A3 is specific for the GC bases in DNA and therefore is an appropriate stain for use in conjunction with Hoechst 33258, as will become evident in the discussion of chromosome techniques. Propidium iodide, although not very specific (it stains all double-stranded regions of both DNA and RNA by intercalating between the stacked bases of the double helix) and not able to penetrate an intact cell membrane, has the decided advantage of absorbing 488 nm light and then fluorescing at wavelengths above 600 nm. This means that, if used in the presence of RNAase, propidium iodide can be used as a DNA stain in cytometers with low-power argon lasers. Propidium

**TABLE 7.1. Characteristics of Some Nucleic Acid Stains**

| Stains | Specificity | Absorption | Fluorescence |
|---|---|---|---|
| HOECHST 33342 and HOECHST 33258 | DNA (with AT preference) | UV | Blue |
| DAPI | DNA (with AT preference) | UV | Blue |
| Mithramycin and chromomycin A3 | DNA (with GC preference) | Blue | Blue-green |
| Propidium iodide and ethidium bromide | Double-stranded nucleic acids | Blue-green | Red |
| Acridine orange | Nucleic acids | Blue-green | Green for DNA and red for RNA |
| Pyronin Y | RNA | Blue-green | Red |
| Thiazole orange | RNA | Blue-green | Green |

iodide has therefore become the most common DNA fluorochrome for flow analysis. Other fluorochromes that absorb 488 nm light include acridine orange, which is metachromatic; that is, it fluoresces red if bound to nonhelical nucleic acid (*e.g.*, RNA or denatured DNA) and fluoresces green if bound to helical nucleic acid (*e.g.*, native DNA). Acridine orange has been used effectively by Darzynkiewicz and coworkers to follow the changes in RNA content and in DNA denaturability that occur during the cell cycle. Moreover, the monoclonal antibody against bromodeoxyuridine (a thymidine analog) can be conjugated to fluorescein and it will then stain DNA that has incorporated bromodeoxyuridine when cells have been pulse-fed with this compound during DNA synthesis. Before discussing the uses of these stains for chromosome and cell cycle analysis, I will first consider the most obvious use of DNA fluorochromes: staining cells for their total DNA content.

## PLOIDY

The amount of DNA in the nucleus of a cell (called the *2C* or *diploid* amount of DNA) is specific to the type of organism in question. Different species have different amounts of DNA in their cells (*e.g.*, human cells contain about 6 pg of DNA per nucleus; chicken cells, about 2.5 pg of DNA per nucleus; corn [*Zea mays*] nuclei, about 15 pg; and *Escherichia coli*, between 0.01 and 0.02 pg each). However, within the animal kingdom, with two major exceptions, all normal cells in an organism contain the same amount of DNA. (The two exceptions are, first, cells that have undergone meiosis in preparation for sexual reproduction and therefore contain the 1C or haploid amount of DNA typical of a gamete; and, second, cells that are carrying out DNA synthesis in preparation for cell division (mitosis) and therefore for a short period contain between the 2C amount of DNA and twice that amount.) Because normal cells, with these two exceptions, contain the same amount of DNA, measurement of the DNA content of cells can be used to identify certain forms of abnormality. More specifically, the type of abnormality commonly termed *malignancy* is often associated with genetic changes, and these genetic changes may sometimes be reflected in changes in total DNA content of the malignant cell.

It is feasible to permeabilize the outer membrane of normal cells (with detergent or alcohol) in order to allow propidium iodide to enter the nuclei. If we then treat the normal cells with RNAase in order to ensure that any fluorescence results from the DNA content (and not the double-stranded RNA content) of the nuclei, we find that the nuclei fluoresce red with an intensity that is more or less proportional to their DNA content. By the use of a red filter and a linear amplifier on the photomultiplier tube, we can detect that red fluorescence; the channel number of the fluorescence intensity will be proportional to the DNA content of the cells. The method is simple and takes about 10 minutes. Flow cytometric analysis of the red fluorescence (605 nm) from the particles in this preparation of nuclei from normal, nondividing cells will result in a histogram with a single, very narrow peak (see the first histogram in Fig. 7.1); all the particles emit very nearly the same amount of red fluorescence. This supports our knowledge that all normal nuclei from any one organism contain the same amount of DNA.

If we then look at a preparation of material from malignant tissue, we find that the fluorescence histogram often indicates the presence of particles with the "wrong" amount of DNA, as well as particles with the amount of DNA that is normal for the organism in question. The normal cells are said to be *euploid* or normal diploid, and the abnormal cells are termed *aneuploid* (flow cytometrists have hijacked these terms from the cytologists and use them to refer to total DNA content of cells; cytologists quite correctly feel that the use of the euploid/aneuploid classification is ambiguous unless chromosomes have been counted). Histograms from

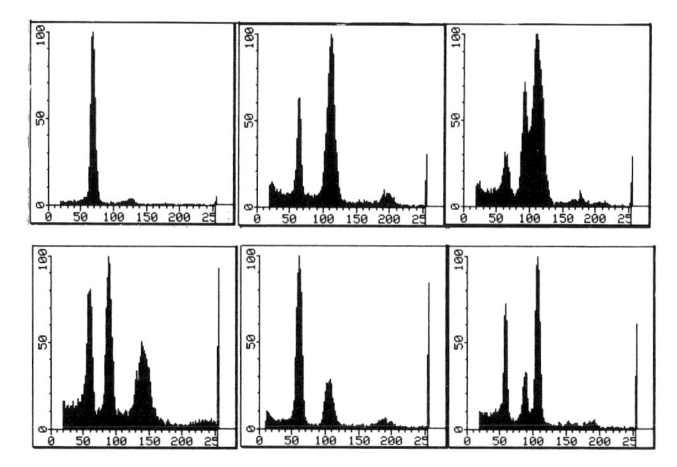


**Fig. 7.1.** Propidium iodide fluorescence histograms from nuclei of cells aspirated from normal tissue and malignant breast tumors. Data courtesy of Colm Hennessy.

examples of some malignant tissues are shown in Figure 7.1. The abnormal peak or peaks may have more or less DNA than normal cells (hyperdiploid or hypodiploid). Because our basic axiom is that all normal cells from an organism contain the same amount of DNA, any tissue that yields a DNA flow histogram with more than one peak contains, by definition, abnormal cells. Flow cytometry is therefore a quick and straightforward method for measuring the presence of the particular type of abnormality that results in cells with abnormal DNA content.

About 5 years ago, at the same time that scientists were beginning to realize that changes in the total DNA per nucleus could be measured by flow cytometry and that this could be an indicator of the presence of abnormal tissue, David Hedley in Australia discovered that when fixed material embedded in paraffin blocks was de-waxed amd rehydrated, released nuclei could be analyzed by flow cytometry for DNA content. Although the absolute fluorescence intensity of PI-stained nuclei released from fixed material was lower than from fresh material, the patterns revealed in the flow histograms were similar. The finding that material from paraffin blocks could be used to analyze DNA content (ploidy) of the individual cells had two important consequences. A very large amount of archival clinical material was suddenly amenable to DNA analysis. And because some of the archival material was 5, 10, and 20 years old, long-term clinical follow up on the patients was immediately available.

The correlation of DNA flow histograms with prognosis became a quick and simple proposition. As a result of Hedley's technique, the corridors in hospitals all over the world were suddenly filled with swarms of ambitious young clinicians beating paths to the doors of their pathology departments (and then on to the flow cytometry facilities). An enormous number of publications emerged, and are still emerging, from the use of this technique on many different types of human material. Aware of the risk of over-generalization and without the time or space in this book for a full discussion of clinical correlations, I can probably safely say here that most (but not all) of the published results show correlation between abnormal flow histograms (aneuploidy) and unfavorable prognosis. And many of the publications also show that flow histograms provide information about prognosis over and above that provided by other, more traditional prognostic indicators.

Although the use of DNA flow histograms to diagnose malignancy is both easy and rapid, it does have certain drawbacks that should be made clear. The first results from the nature of the malignant changes themselves: Not all malignancies will result from DNA changes that are detectable by a flow cytometer. Knowledge of the causes of malignancy is far from perfect. Nevertheless, it is possible to imagine that some malignancies may result from changes that are not related to DNA, and other malignant changes may affect a cell's DNA but not in a way that could ever be detected in a flow histogram of propidium iodide fluorescence. For example, chromosome translocations may lead to gross abnormalities in genetic coding, but do not lead to any change at all in the total DNA content of a nucleus. Translocations can be detected easily by microscopic analysis of the individual banded chromosomes in a mitotic spread, but translocations will never be detected by flow cytometry of nuclei stained with propidium iodide. Similarly, small insertions or deletions to chromosomes may lead to changes in DNA content that are too small to be detected by flow cytometry (any change resulting in less than 5% difference in total DNA content may be difficult to detect). Whereas extra copies (*e.g.*, trisomy) of a large chromosome may result in a measurable shift in the total DNA content of a nucleus, trisomy of a small chromosome may not be detectable in this type of flow analysis (a large chromosome might contain 4% of a cell's total DNA, but a small chromosome has less than 1%).

While this first kind of difficulty is an intrinsic limitation of the DNA flow histogram technique, a second problem is more in the nature of a continuing question about interpretation. Although Figure 7.1 shows examples of histograms that provide undoubted evidence of abnormality, Figures 7.2 and 7.3 show another series of histograms that are considerably more difficult to interpret. Figure 7.2 shows the problems that arise with so-called wide coefficient of variation (CV) data; the real question is our ability to rule out the existence of near-diploid abnormalities when the width (CV) of a peak is very broad. In theory, since all normal nuclei contain the same amount of DNA, the peak in a flow histogram of normal cells should have the width of only a single channel (all the particles should have the same fluorescence intensity and should appear in the same channel). In practice, because staining and illumination conditions may not be exactly uniform, the fluorescence intensity of normal nuclei stained with propidium iodide will have a certain range of values. One of the ways in which otherwise quite civilized flow cytometrists compete with each other is by bragging about the small CVs on the peaks of their DNA histograms (as a rule of thumb, 1%–2% is good; >8% is not good). A wide CV might result from old and partially degraded material, from a stream and laser beam poorly aligned in relation to each other, from a sample that has been run too quickly (remember that widening of the core diameter within the sheath stream may

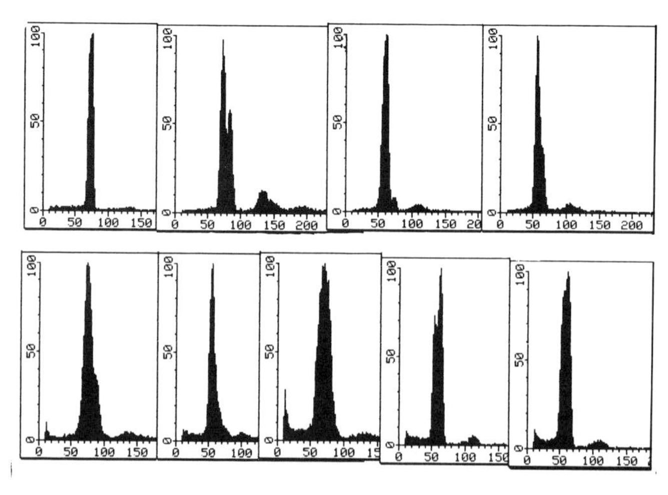


**Fig. 7.2.** Compared with the narrow peak in the normal histogram at the upper left, it can be seen that a single peak with a wide CV or skewed profile may mask a near-diploid malignant cell line. Data courtesy of Colm Hennessy.

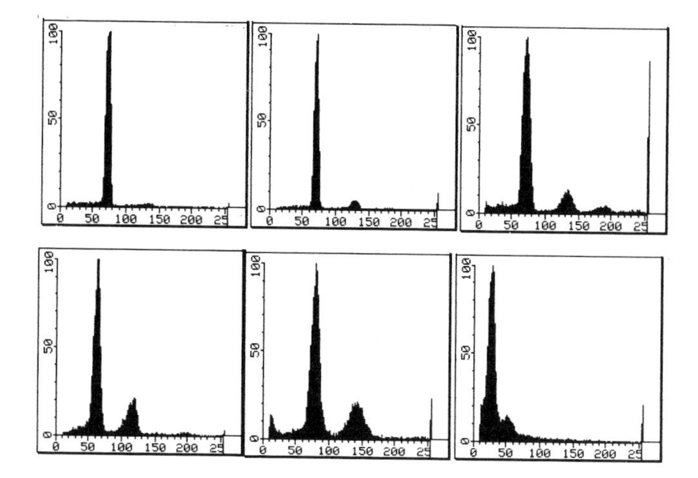


**Fig. 7.3.** A peak at twice the normal diploid amount of DNA may result from clumping of nuclei, cycling cells, or a true tetraploid abnormality. These three possibilities are not always readily distinguished from each other. Data courtesy of Colm Hennessy.

lead to unequal illumination as particles stray from the center of the laser beam), or, finally, from abnormal cells with a DNA content quite close to that of the normal material. The sensitivity of the technique for detecting these near diploid abnormalities and thus for classifying tissue as euploid or aneuploid therefore depends on the cytometrist's ability to obtain narrow CVs in the normal controls.

The second problem concerning interpretation arises from the inconvenient fact that aneuploid tumors often have DNA content that is very close to double the amount found in normal cells. This amount is referred to as *4C* or *tetraploid* (Fig. 7.3). If we stop and think about this fact, we can immediately see why this might lead to problems in flow analysis. First of all, perfectly normal cells with the 4C amount of DNA appear at certain phases in the cell cycle (just before cell division); therefore, if normal dividing cells are present, a significant number of particles may have double the 2C amount of DNA and will therefore appear in a peak at the tetraploid position. Second, remember that the flow cytometer is especially poor in its ability to distinquish large particles from clumped particles. It is not surprising, then, that the cytometer is, in the same way, inadequate at distinguishing a nucleus with double the normal amount of DNA from two normal nuclei clumped together. Cells in the 4C position as a result of impending mitosis can, to a certain extent, be defined by other characteristics that one might expect from a cycling population. Clumped cells, however, are not so easy to distinguish from an abnormal tetraploid cell line. One way of diagnosing a clumped sample is by looking for peaks at the 6C position (resulting from three nuclei together).

By making use of so-called pulse processing technology, we can get more help in this task. Pulse processing involves analysis of the time profile of the fluorescent signal from a particle. Because one large nucleus will pass through the laser beam more quickly than will two smaller nuclei of the same total volume, the time characteristics of the resulting fluorescence signal can be used to distinquish clumps from single particles and can help in interpretation of DNA profiles with peaks in the tetraploid position.

Nevertheless, histogram peaks at the double diploid position remain difficult to interpret because of these two forms of ambiguity. Scientists and clinicians usually resort to adopting some threshold value for classification purposes. For example, a sample with a 4C peak may be considered aneuploid only if the tetraploid peak contains more that 10% (or 15%) of the total number of nuclei counted; otherwise it will be considered normal on the assumption that about 10% of the particles may normally appear in the tetraploid position owing to clumping and/or mitosis.

## CELL CYCLE ANALYSIS

As mentioned above, normal cells will have more DNA than the 2C amount appropriate to their species at times when they are preparing for cell division. The cell cycle has been divided into phases (Fig. 7.4). Cells designated as being in the G0 phase are not cycling at all; cells in G1 are sitting around either just recovering from division or preparing for the initiation of another cycle; cells are said to be in S phase when they are actually in the process of making new DNA; cells in the G2 phase are those that have finished DNA synthesis and therefore possess double the normal amount of DNA; and cells in M phase are undergoing the chromosome condensation and organization of mitosis that occur immediately prior to cytokinesis with the production of two daughter cells, each with the 2C amount of DNA. A DNA flow histogram provides a snapshot of the distribution of different kinds of nuclei present at a particular moment. If we look at the DNA content of cells that are dividing, we will find some

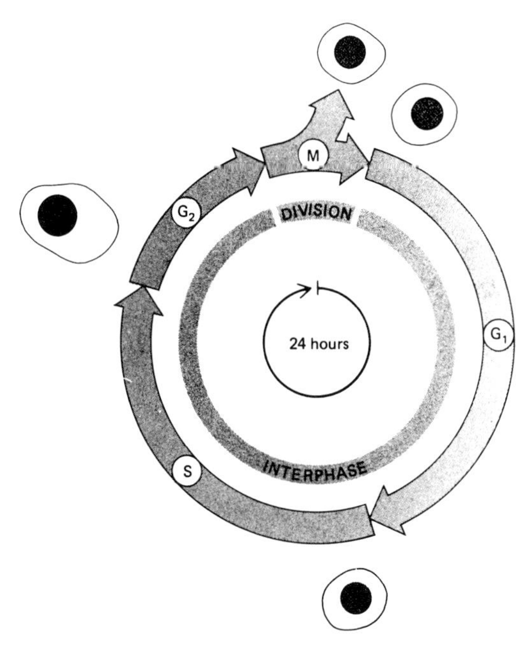


**Fig. 7.4.** The four successive phases of a typical eukaryotic cell cycle. From Alberts et al. (1989).

nuclei with the 2C amount of DNA (either G0 or G1 cells), some nuclei with the 4C amount of DNA (G2 or M cells), and some nuclei with different amounts of DNA that span the range between these 2C and 4C populations (Fig. 7.5). A theoretical histogram distribution would look like Figure 7.6. Figure 7.7 shows actual examples of DNA flow histograms that result from the propidium iodide staining of cells taken at intervals from a culture after it has been stimulated to divide.

The traditional method for analyzing cell division involves measuring the amount of DNA being synthesized in a culture by counting the radioactivity incorporated into DNA when the dividing cells are given a 6 hour pulse with tritiated thymidine. The DNA histogram resulting from flow cytometric analysis offers an alternative to this technique. By dividing the histogram up with four markers, we can delineate nuclei with the 2C amount of DNA, those with the 4C amount of DNA, and those with amounts of DNA between the two delineated regions and therefore caught

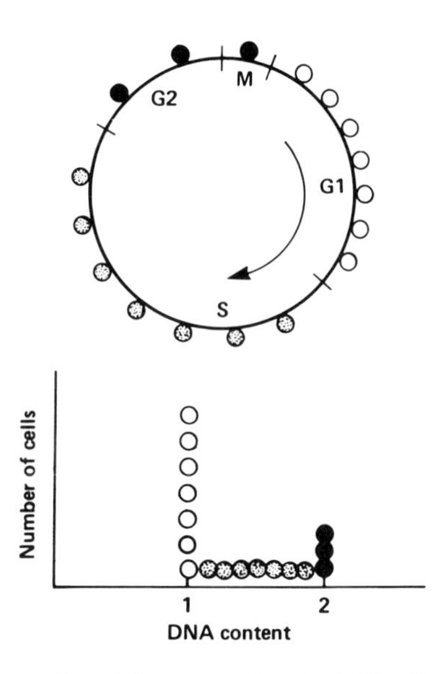


**Fig. 7.5.** Schematic illustration of the generation of a DNA distribution from a cycling population of cells. From JW Gray et al. (1990).

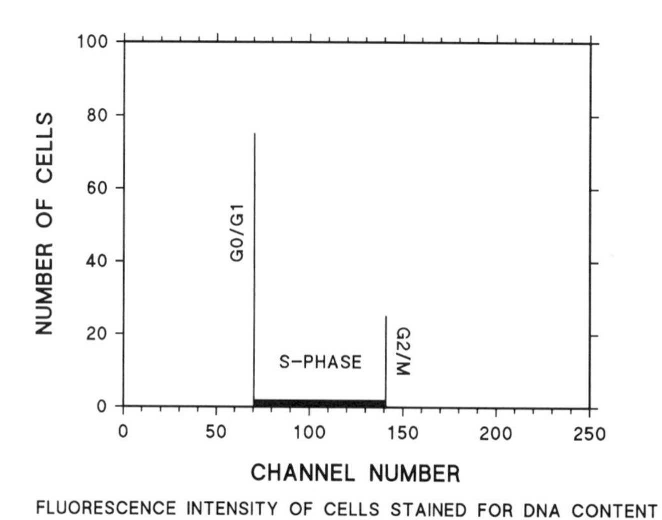


**Fig. 7.6.** The theoretical histogram generated from the sampling of a population of cycling cells.

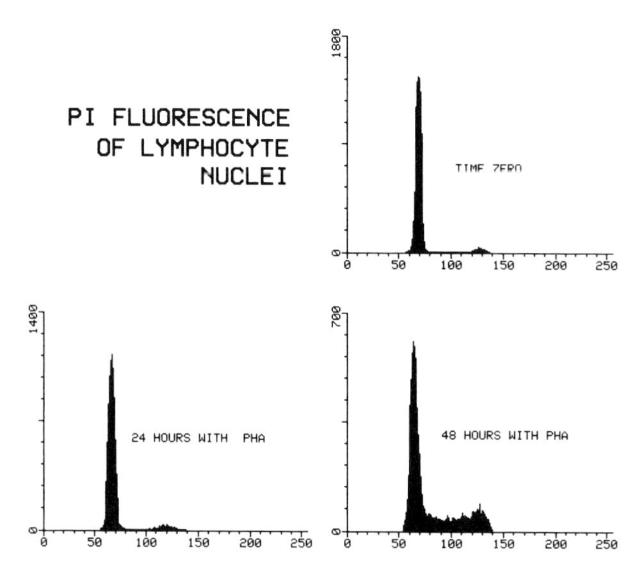


**Fig. 7.7.** DNA histograms from lymphocytes stimulated to divide. Data courtesy of Ian Brotherick.

in the process of synthesizing DNA. The nuclei making DNA and showing up between the two peak regions should in some way correlate with the values obtained for DNA synthesis based on the uptake of tritiated thymidine. The values are not directly convertible one to the other; the radioactive method measures the total amount of DNA being synthesized and will give higher values when more cells are present, whereas the flow method measures the percentage of cells that are in the process of making DNA and will not be affected by increases in the total number of cells. In addition, the radioactive method will give higher values if there is a significant amount of DNA repair going on, whereas the flow method will give higher values if a proportion of cells are blocked in S phase. However, with these provisos, flow cytometry does offer a rapid and painless (nonradioactive) method for looking at cell division.

Having agreed on the general principle that flow cytometry in conjunction with propidium iodide staining is an appropriate technology for analyzing cell division, we have now to face the problem that the actual histogram does not look like our theoretical distribution; we have to decide where to place those four markers mentioned above so as to delineate correctly the three regions (2C, 4C, and S phase). In a scenario that may by now be familiar, what seemed like a straightforward question turns out to have a less than straightforward answer. Because the 2C and 4C peaks in a flow histogram have finite widths (remember the discussion about CV in the section on ploidy), it turns out to be rather difficult to decide where the 2C (or G0/G1) peak ends and nuclei in S phase begin. Similarly, it is difficult to know exactly where the distribution from nuclei in S phase ends and the spread from nuclei in G2 or M (4C amount of DNA) begins. In fact, there is no unambiguously correct point to place markers separating these three regions: The regions overlap at their extremes as a result of the inevitable non-uniformity of staining and illumination. The question therefore becomes not where to place the markers delineating the three cell cycle regions, but how many of the nuclei lurking under the normal spread of the 2C and 4C regions of the histogram are actually in S phase. Enter the mathematicians.

Algorithms based on sets of assumptions about the kinetics of cell division and the resulting shape of cell cycle histograms can be used to derive formulae for separating the contribution to the fluorescence distribution from our three separate cell cycle components. The algorithms used range from the simple to the complex, and new refinements are being proposed all the time. They all seem to work reasonably well (that is, they all give similar and intuitively appropriate answers) when cell populations are well behaved. However, they all reflect the intrinsic limitations of using simplistic mathematical models for complex biological systems when cell populations grow too rapidly, are blocked in the cycle, or are otherwise perturbed. Bearing these limitations in mind, we may now look at four of the models used.

Figure 7.8 indicates a DNA histogram derived from the propidium iodide staining of cells from a dividing culture. The simplest method for analyzing this histogram is the so-called *peak reflect method* whereby the shape of the G0/G1 peak is assumed to be symmetrically distributed around the mode. Given this assumption, the width of the peak, from the mode to the left (low fluorescence) edge is simply copied to the right (high fluorescence) edge; the same thing is done in reverse with the G2/M peak. Then everything in the middle between these two delineated regions is considered to be the result of S phase cells.

A slightly more complex method for estimating the proportion of S phase cells is called the *rectangular approximation method.* This method assumes that cells progress regularly through S phase and therefore that the proportion of cells at any given stage of DNA synthesis is constant. When this

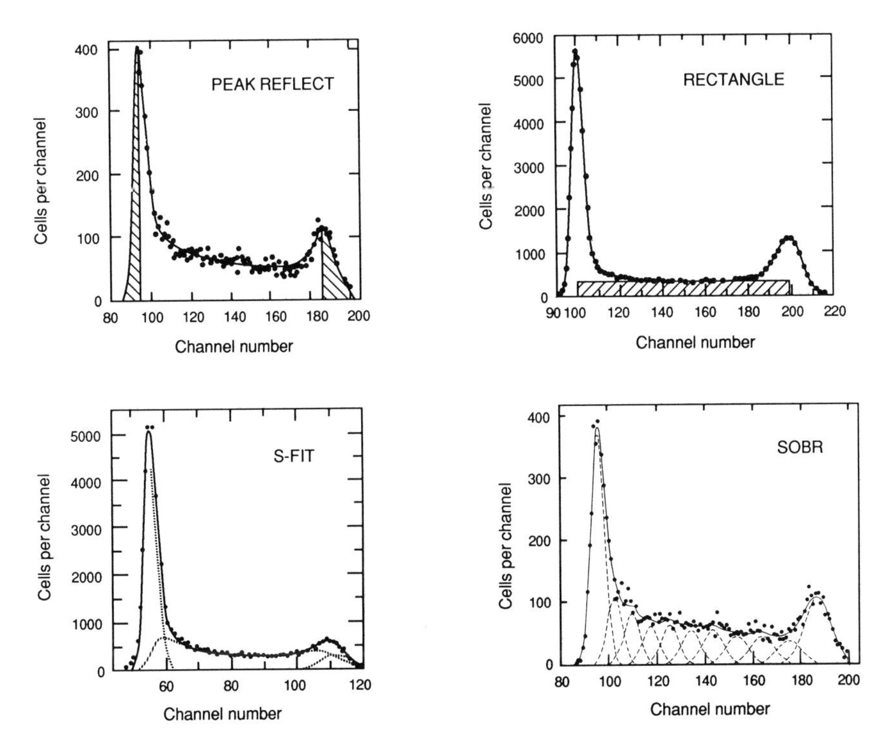


**Fig. 7.8.** Different mathematical algorithms for determining the contribution of S phase nuclei to a DNA flow histogram. A and D from Dean (1987); B and C from Dean (1985).

method is used, the average number of cells in the middle region of the DNA histogram is evaluated, and the height of this region is then extrapolated in both directions, toward the 2C peak and toward the 4C peak. The rectangle derived from this evaluation is then ascribed to S phase cells, and all the other cells are considered either G0/G1 or G2/M depending on whether they have higher or lower fluorescence than the middle point of the distribution.

The so-called S-FIT method and the Sum-of-Broadened-Rectangles (SOBR) method both use more sophisticated mathematical assumptions to model the shape of the S phase region of the histogram. A polynomial equation (S-FIT) or a series of broadened gaussian distributions (SOBR) is derived that best fits the S phase region of the histogram; then this derived shape is extrapolated toward the 2C and 4C peaks to estimate the contribution of S phase cells within these regions.

All these methods for estimating the proportion of S phase cells are rough approximations. They are being continually refined mathematically in the hope of better approaching the biological truth; but a mathematical model will never be able to cope well with a biological situation that is disturbed or contains mixed populations behaving in erratic ways. Flow cytometry does, however, offer a more direct way to measure DNA synthesis. The bromodeoxyuridine method is, in fact, directly comparable to the traditional method of measuring the incorporation of tritiated thymidine.

Bromodeoxyuridine (sometimes abbreviated BrdU or BUdR or BrdUdr) is a thymidine analog; if cells are pulsed with BrdU, it will be incorporated into the cell's DNA in the place of thymidine. FITC-conjugated monoclonal antibodies with specificity for BrdU are available, so that cells that have been pulsed with BrdU for a short period of time (about 30 minutes) can then be treated to denature their DNA, exposing the BrdU within the double helix so that it can be stained with the anti-BrdU antibody. Any cells that have incorporated BrdU during the pulse will then stain FITC positive. The cleverest part of this technique is that the denatured DNA can be stained with propidium iodide at the same time. The resulting two-color contour plots look like those in Figure 7.9. The red fluorescence axis shows the propidium iodide distribution (proportional to DNA content) with which we have grown familiar; the green fluorescence axis shows which of these nuclei have actually incorporated BrdU during the pulse. As might be expected, the cells in the middle region of the propidium iodide distribution have all incorporated BrdU; but a proportion of the cells at either end of the propidium iodide distribution have also done so. This method, while somewhat time consuming and a bit tricky technically, does allow a flow cytometrist to quantify the proportion of cells in S phase in a way that cannot be done accurately with simple propidium iodide staining.

BrdU staining has also been used to provide information about the kinetics of cell cycles. If we consider cells pulsed for a short period of time with a small amount of BrdU and then killed immediately and stained with both propidium iodide and with the FITC–anti-BrdU monoclonal antibody, the FITC label should stain equally all cells in S phase (stretching from the 2C peak to the 4C peak). If, however, we wait for some time before killing the cells (and if the BrdU has been used up quickly), then the cells that have incorporated the BrdU (that is, all the cells that were in S phase at the time of the pulse) will have synthesized more DNA and some of those cells will

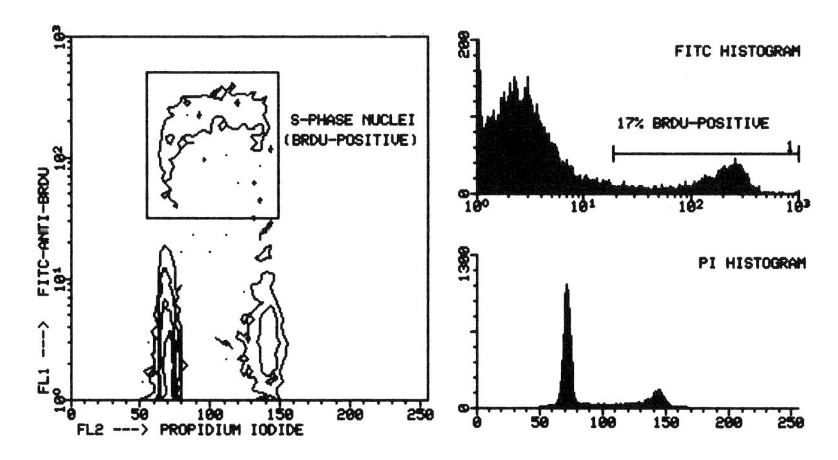


Fig. 7.9. FITC histogram, propidium iodide (PI) histogram, and dual-color correlated contour plot of human keratinocytes cultured for 4 days, pulsed with BrdU, and then stained with FITC-anti-BrdU and PI. Data courtesy of Malcolm Reed.

have progressed into the G2 or M phase of the cell cycle (or indeed cycled back to G1). In addition, some new cells will have started to make DNA after the BrdU had been used up, and these cells will now be in S phase but will have DNA not containing BrdU. The contour plots for these cases look like those in Figure 7.10. We can estimate the rate of movement of the BrdU-containing cells through S phase and into the G2 peak by assuming that they are evenly distributed throughout S phase at the time of pulsing and then sampling and staining the cells at one subsequent time. The rate of increase in propidium iodide intensity of the FITC-positive nuclei is equivalent to their rate of DNA synthesis and provides us with information about the cycle time of the actively dividing cells. Moreover, the cycle time of the FITC-positive cells, in conjunction with the proportion of cells in S phase, can be used to estimate the doubling time of a population of cells.

Although flow cytometrists tend to have reservations about the reproducibility of S-phase determinations from mathematical models of the propidium iodide histogram (for the reasons mentioned above), reports are being published in the medical literature about the correlation of S-phase fraction with clinical prognosis. A more precise, but more difficult, method involves injecting patients with BrdU; their tumors are removed sometime later in order to estimate the potential doubling time of the tumor and, by implication, its aggressiveness.

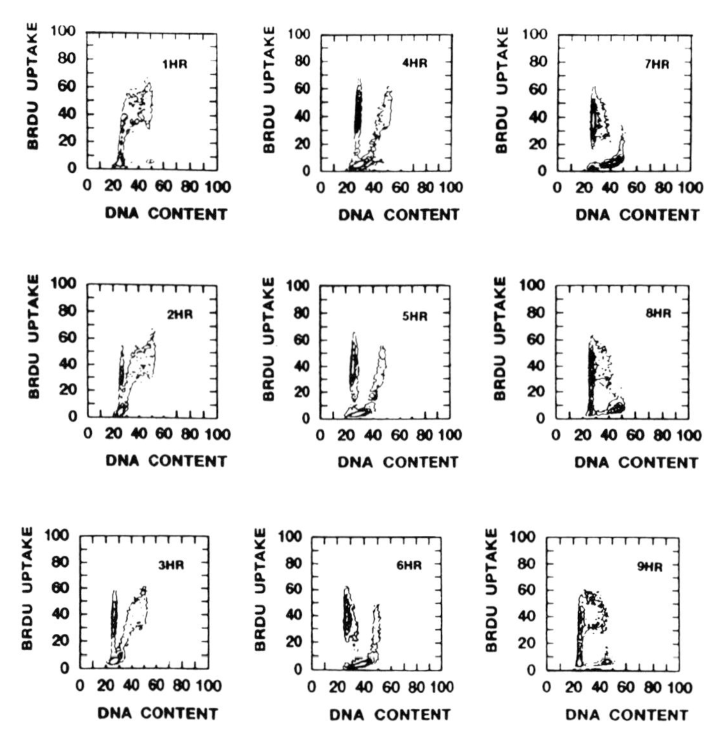


**Fig. 7.10.** Dual-color distribution of BrdU *vs.* DNA content for Chinese hamster cells pulsed with BrdU and then sampled hourly. From McNally and Wilson (1990).

## TWO-COLOR ANALYSIS FOR DNA AND ANOTHER PARAMETER

Having described the analysis of cells for total DNA content (propidium iodide) and newly synthesized DNA (BrdU) simultaneously, we may now go on to consider some other techniques for exploiting the potential of the flow cytometer by the dual staining of cells for both DNA and some other parameter. Darzynkiewicz and coworkers in New York have used acridine orange with great success to look at DNA and RNA contents of cycling cells. Because acridine orange fluoresces red when it binds to RNA and green when it binds to DNA, cells can be examined simultaneously for both constituents. Dual analysis of this type has given us increased information about the progression of cells through the cell cycle. Specifically, it can be seen quite clearly from plots like those in Figure 7.11 that some of the cells

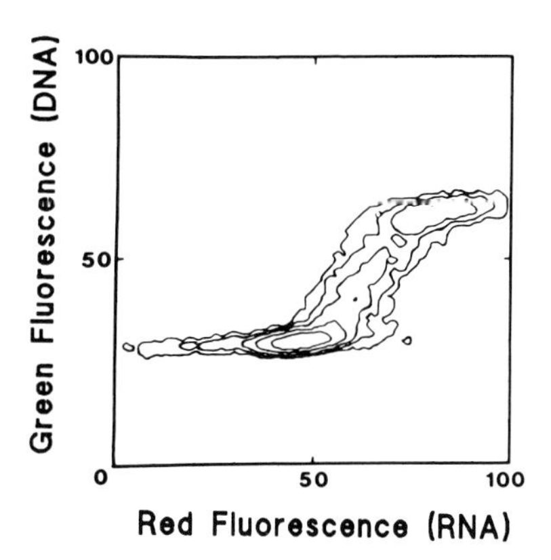


**Fig. 7.11.** DNA (green fluorescence) *vs.* RNA (red fluorescence) content of human leukemic cells stained with acridine orange. From Darzynkiewicz and Traganos (1990).

with the 2C amount of DNA (G0 or G1) contain increased levels of RNA. The synthesis of RNA appears to be an early event in the entrance of a cell into the division cycle. After this initial increase in RNA content, cells synthesize more RNA as they begin to synthesize DNA (S phase). At mitosis, they have increased levels of both RNA and DNA compared with resting cells. Analysis of the acridine orange staining of cycling cells has led, in particular, to improved ability to analyze early events in the division cycle.

This acridine orange technique has been taken one step further by Darzynkiewicz's group. Because acridine orange fluoresces red with single-stranded nucleic acids, but green with double-stranded nucleic acid, acridine orange can be used under mildly denaturing conditions (and in the presence of RNAase) to distinguish easily denatured from more resistant forms of DNA. Once RNAase has been used to remove the RNA, mild denaturation will cause unwinding of the double helix of forms of DNA that are loosely packed (*e.g.*, DNA in cells that are in the process of DNA synthesis), but will not affect DNA that is tightly condensed (*e.g.*, resting cells). An "alpha T ratio" is defined as the ratio of red fluorescence to red plus green fluorescence and is related to the percentage of DNA that is denatured under standard mild denaturing conditions. This ratio has been found to be high in certain kinds of cells, for example, in advanced stage, invasive carcinoma. Figure 7.12 indicates the kind of contour plots that can be obtained from this type of procedure.

Dual staining of cells can also be used to look at DNA and protein markers simultaneously. Various techniques have been developed by which the investigator can first stain the surface markers on the outer membrane of cells, then permeabilize the cells, and finally stain the DNA of the cells with propidium iodide. Figure 7.13 shows the way that this type of technique can be used to permit the cell cycle analysis of subpopulations of cells independently. In this particular example, it can be seen that, after treatment with the mitogen phytohemagglutinin, it is the $CD8^+$ cells more than the $CD8^-$ cells that have been induced to enter S phase.

Similar techniques can be used to stain cells for both DNA and internal (cytoplasmic or nuclear) markers. By fixing the cells, then staining for the cytoplasmic markers (like cytokeratin, which stains epithelial cells only), and then staining with propidium iodide, classes of cells can be selected for ploidy analysis. In this way, for example, breast tumor cells (which are likely to be of epithelial origin), can be gated in a mixed population from a tumor. Then the cells gated for FITC-anti-cytokeratin positivity can be further analyzed to see if any of the nuclei are aneuploid. By this technique,

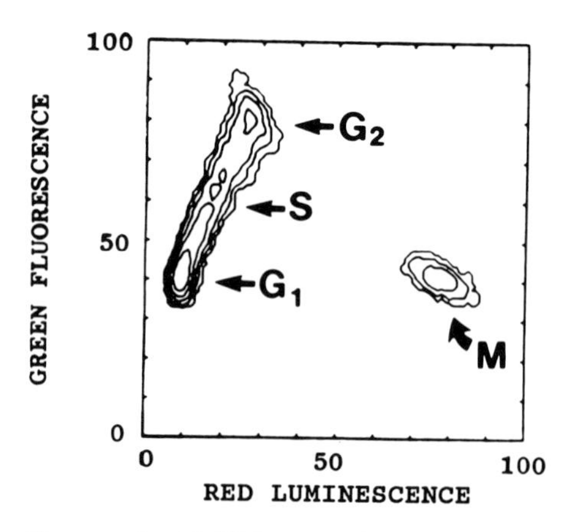


**Fig. 7.12.** L1210 cells treated with RNAase and acid and then stained with acridine orange reveal that DNA in mitotic cells is extensively denatured and exhibits increased red and decreased green fluorescence relative to DNA from cells in other phases of the cell cycle. From Darzynkiewicz (1990).

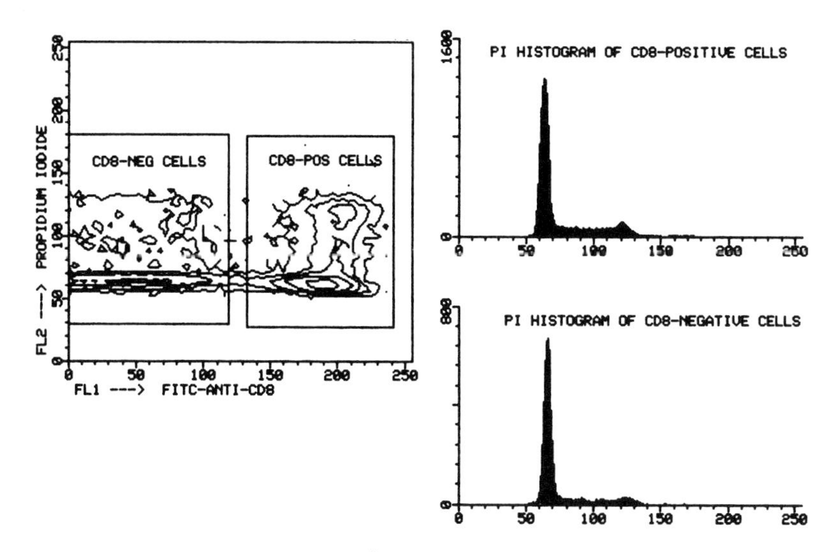


**Fig. 7.13.** Cell cycle analysis of CD8$^+$ and CD8$^-$ cells. Lymphocytes were cultured with PHA, stained with FITC-anti-CD8 monoclonal antibody, treated with saponin to permeabilize the outer membrane, and then stained with PI and RNAase. Cells provided by Ian Brotherick.

minor populations of tumor cells within a heterogeneous population may be detected and their ploidy determined.

## CHROMOSOMES

Until now, the particles flowing through our flow cytometer have been cells or nuclei. But I have tried to avoid use of the word *cell* when describing the technical aspects of flow cytometry as a reminder that other types of particles can flow through and be analyzed by a cytometer. One of the best examples of the successful application of flow cytometry to noncellular systems has been in the analysis of chromosomes. In this case, the particles flowing through the system are individual chromosomes that are released from cells that have been arrested in metaphase (much the same conditions as those that are used to prepare chromosomes for analysis in metaphase spreads under the microscope). The released chromosomes are stained with a DNA stain (like propidium iodide) and then sent through the flow cytometer; the resulting histograms of fluorescence intensity reveal peaks whose position along the x-axis is proportional to the amount of DNA in the chromosome and whose area is proportional to the number of chromosomes with that particular DNA content (Fig. 7.14). Histograms of this type are called *flow karyotypes*, by analogy with the microscope karyotypes derived from conventional genetic analysis. Figure 7.15 shows examples of flow karyotypes from different species. While some species with small numbers of chromosomes reveal relatively simple histogram patterns, the 23 pairs of chromosome in the human lead to a rather complex pattern. It is apparent that, although 23 pairs of chromosomes are readily distinguished under the microscope by a combination of size, centromere position, and banding patterns, many of these pairs have similar total DNA content and are not distinguishable in a flow histogram. We can obtain considerable help by using Hoechst 33258 and chromomycin A3 in a dual staining system: Hoechst 33258 stains adenine- and thymine-rich regions

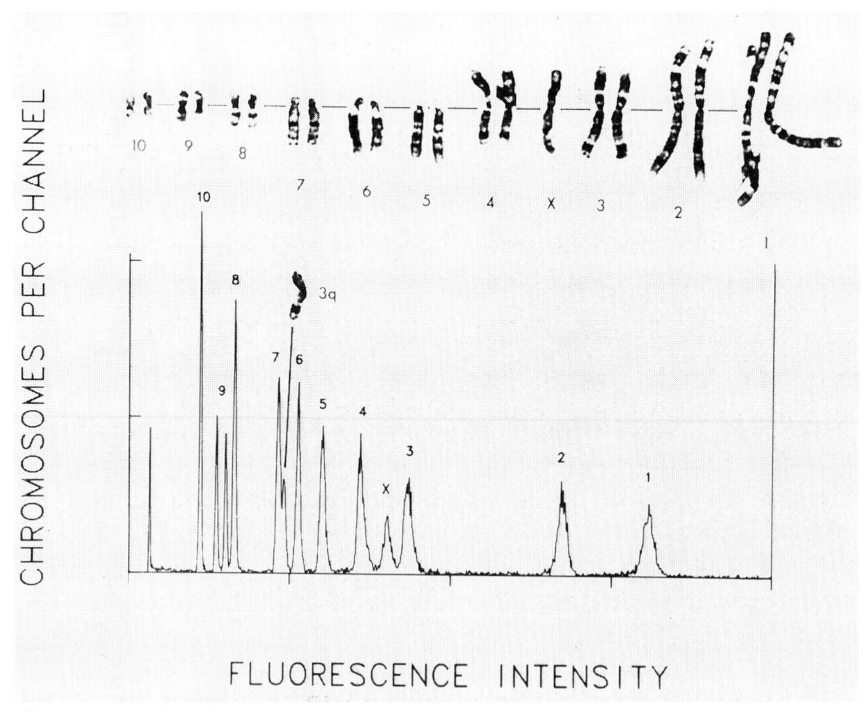


**Fig. 7.14.** A flow karyotype (fluorescence histogram) of Chinese hamster chromosomes stained with PI. The G-banded chromosomes from this particular aneuploid cell line are included for comparison with the histogram peaks. From Cram et al (1988).

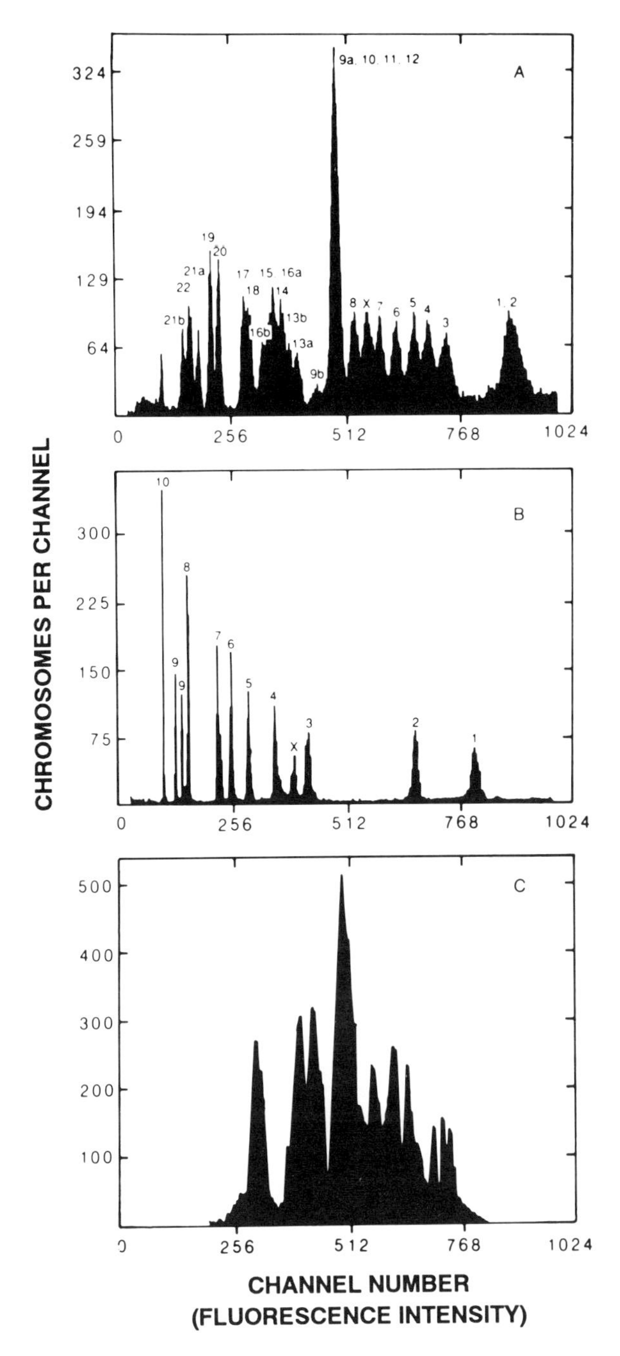


**Fig. 7.15.** Flow karyotypes from human chromosomes (**A**), hamster chromosomes (**B**), and mouse chromosomes (**C**). From Gray and Cram (1990).

of DNA preferentially, and chromomycin A3 is specific for regions rich in the guanine and cytosine base pairs. By using this dual system, we find that some chromosomes with closely similar total DNA contents have differing base pair ratios. Compare particularly the positions of chromosomes 13–16 in the one-dimensional histograms with these same chromosomes (now separable) in the contour plot (Fig. 7.16). Unfortunately, chromosomes 9–12 are not distinguishable with either system.

There has been much discussion about the potential utility of flow cytometry of chromosomes for clinical diagnosis. As regards its sensitivity, this technique appears to stand somewhere between the technique of flow analysis of whole cells for DNA content and that of microscope analysis of banded chromosomes. It may be a useful intellectual exercise for readers to ask themselves which technique or techniques would be most appropriate for detecting the following types of chromosome abnormality: 1) tetraploidy, where the normal chromosome content of cells is exactly doubled because of failure of cytokinesis at one mitosis; 2) an inversion in an arm of one particular chromosome; and 3) trisomy (the existence of cells with three instead of two) of one of the small chromosomes. In addition to these limitations, the use of flow cytometry to look for abnormal chromosomes has been confounded by the fact that several human chromosomes are

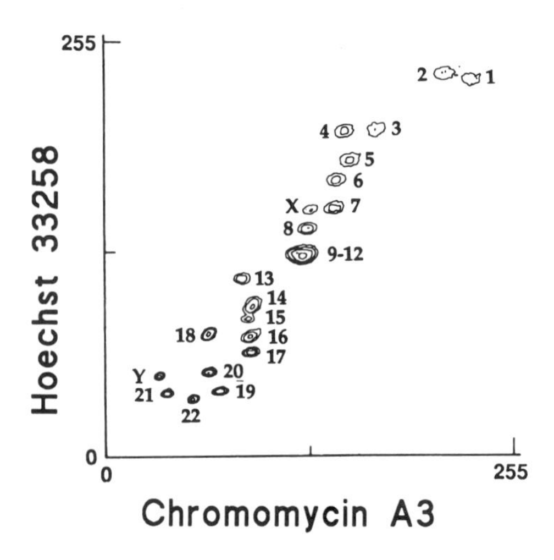


**Fig. 7.16.** A bivariate flow karyotype for human chromosomes stained with Hoechst 33258 and chromomycin A3. From Gray and Cram (1990).

highly polymorphic, and therefore flow karyotypes vary considerably among normal individuals.

More detailed information about chromosomes is, however, being obtained by so-called slit-scan cytometry. In general, particles give out signals that last, in time, just as long as it takes for the particle to move through the laser beam. What this means is that particles with diameters smaller than the beam diameter all give out signals that last approximately the same length of time (dependent primarily on the laser beam width in the direction of flow and on the stream velocity); this is the traditional method of flow analysis (Fig. 7.17). However, it is apparent that larger particles (or small particles in a very narrow laser beam) will give out signals whose time profiles are related primarily to their own diameter. This type of signal analysis is currently being used to distinguish large single nuclei from clumps of nuclei in DNA analysis; two clumped nuclei take longer to pass through the laser beam than does one large one of an equivalent total volume. It is being carried a step further by the design of so-called slit-scan instrumentation.

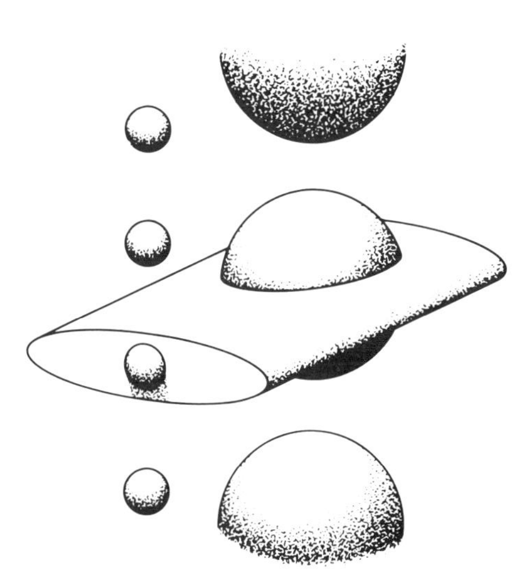

**Fig. 7.17.** The time that a fluorescence signal lasts depends on the size of the laser beam (if the cell is smaller than the beam) or on the size of the cell (if the cell is larger than the beam). From Peeters et al. (1989).

With slit-scan technology, the laser beam is narrowed to about 0.7μm in the direction of particle flow. This means that even rather small particles, instead of giving out one signal related to their total fluorescence while in the stream, can have their fluorescence sampled at multiple intervals as they pass through the beam. Such slit-scan technology has been finding success in looking at long, thin particles like chromosomes; indeed, the time profile of the propidium iodide fluorescence from a chromosome can be used to look for centromere regions (lower DNA content) and can also be used to look for dicentric chromosomes that result from radiation damage (Figs. 7.18 and 7.19). These techniques clearly allow more detailed analysis of chromosome structure. But there are reservations about whether this kind of flow procedure will ever be as good as the traditional microscopic methods, which are now being enhanced by the techniques of image cytometry and *in situ* hybridization in conjunction with computer recognition systems. However, the discussion of chromosomes leads into the field

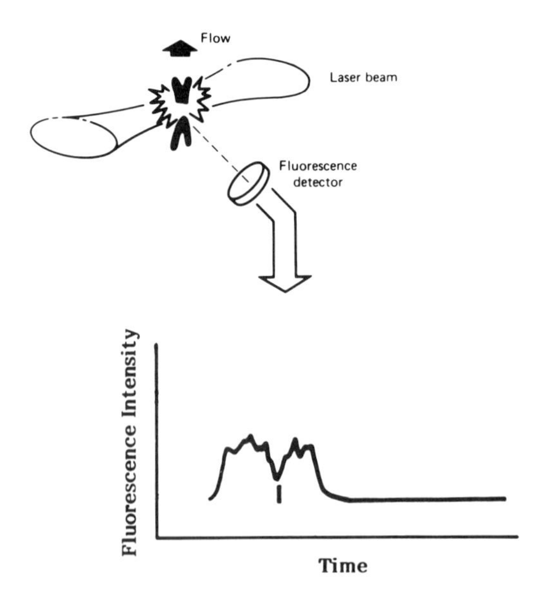


**Fig. 7.18.** Slit-scan flow cytometry of a chromosome. The PI fluorescence intensity coming from the narrow analysis point is measured periodically (*e.g.*, every 10 nanoseconds) revealing a dip in DNA concentration in the region of the centromere. From Gray and Cram (1990).

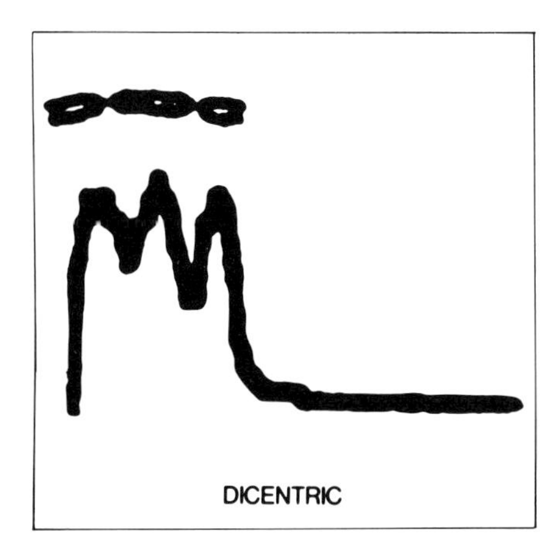


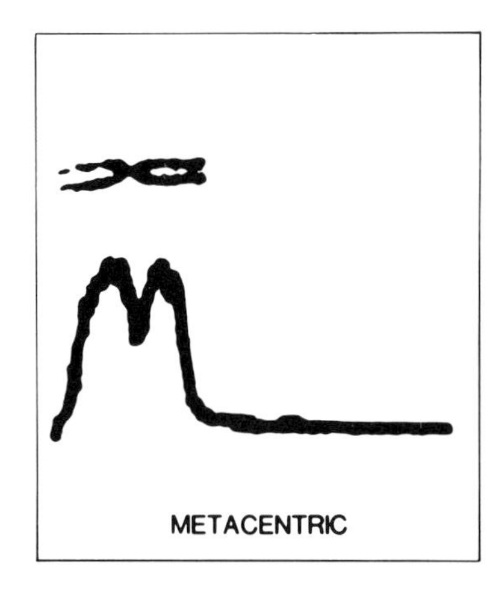


**Fig. 7.19.** Slit-scan flow cytometry can reveal abnormal dicentric chromosomes. From Gray and Cram (1990).

of molecular biology, where interactions with flow cytometry are proving to be both challenging and productive.

## MOLECULAR BIOLOGY

Flow cytometry of chromosomes does have a very definite role to play in the effort toward goals like the formation of DNA libraries and the mapping of the human genome. This is true because of the ability of flow cytometers to sort. Flow sorting, based on the Hoechst 33258 and chromomycin A fluorescence of chromosomes, turns out to be one of the best ways available for obtaining relatively pure preparations of each type of chromosome. Even those chromosomes that are not distinguishable by their fluorescence (*e.g.*, 9–12) can usually be sorted from hamster–human hybrid cell lines. And these preparations of reasonably pure flow-sorted chromosomes are the starting material for obtaining chromosome-specific DNA. The DNA extracted from the sorted chromosomes can be digested with restriction enzymes and the resulting fragments amplified by growth in a recombinant DNA phage vector on a bacterial lawn. The Human Genome Project is making extensive use of the DNA libraries for each chromosome obtained with the high-speed sorting cytometers at Los Alamos and

Livermore. The bottleneck in this technique is in the time it takes for sorting. In a conventional sorter, if the flow rate is limited to about 1,000 particles per second, then chromosomes of a particular type could theoretically be sorted at a rate of about 40 per second (1,000/23). Thus it would take about 7 hours to obtain $10^6$ chromosomes of a given type under optimal conditions. In practice, the time is closer to 15–20 hours. The high-speed sorters at Los Alamos and Livermore were developed with the demands of the Human Genome Project in mind. They utilize high-pressure systems to sort at approximately 10 times the standard rate. Until now, the DNA libraries obtained from sorted chromosomes have consisted mostly of complete digests (that is, small DNA fragments). The next phase will need greater amounts of chromosome-specific DNA in an attempt to produce large insert libraries to generate a map of the human genome. Advances in molecular cloning techniques such as pulsed-field gel electrophoresis and the polymerase chain reaction have decreased the reliance on flow sorting, but the high-speed sorting facilities may still be an important part of this effort.

Small numbers (20,000 or so) of sorted chromosomes, however, can be used quite neatly to aid in the mapping of genes to chromosomes. Chromosomes of each type are simply sorted (two at a time: one type to the left, the other to the right) onto a nitrocellulose filter. The DNA is then denatured on the filter where it can be hybridized to a radioactive gene probe. Autoradiography of the filter will then reveal whether the probe has hybridized to the DNA from any given chromosome (Fig. 7.20). In this way, the sorting of small numbers of each of the chromosomes onto filters allows the mapping of any available gene probe to its chromosome.

Another application of flow cytometry to molecular biology involves the investigation of gene expression. Particularly in studies of the expression of genes for cell surface molecules, flow sorters can select for cells that express high levels of a gene by sorting for high fluorescence intensity on the surface of stained cells. This technique has been used to select for rare somatic cell genetic variants, for example, to isolate isotype switch variants expressing different immunoglobulin subclasses on their surface. It has also been used to select and enrich for high expressing clones from transfected or hybrid cultures. The correlation of the retention of a particular chromosome in a transfected or hybrid cell with a high level of surface marker density can lead to conclusions about the chromosome location of the gene for the surface antigen even when the gene itself has not been identified. Oncogene expression has also been studied by flow cytometry, where monoclonal antibodies directed against the c-myc protein, for example,

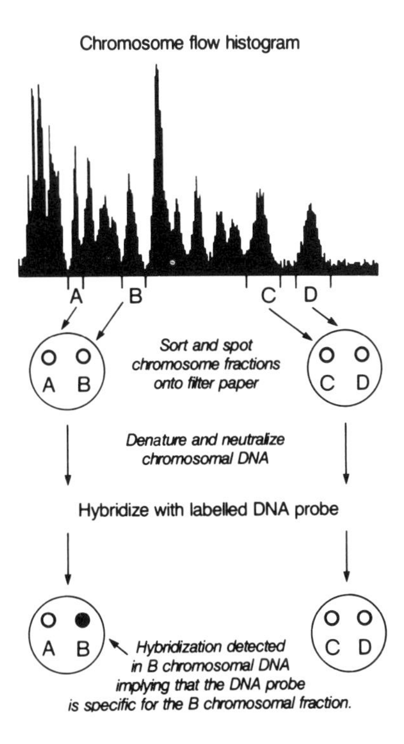


**Fig. 7.20.** DNA blot analysis using spots from chromosomes sorted directly onto filter paper on the basis of their Hoechst 33258 fluorescence. From Van Dilla et al. (1990).

allow investigation of patterns of expression and the associated chromosomal rearrangements.

A recently proposed application of flow cytometry to molecular biology involves use of cytometers that have been developed by Jett at Los Alamos with vastly increased fluorescence sensitivity. These cytometers are able to detect the fluorescence from single molecules. It has been suggested that this capability could be applied to the problem of DNA sequencing. The proposed technique involves the synthesis of a complementary strand of DNA with fluorescently tagged precursors (each base of a different color). The labeled (fluorescent) duplex DNA molecule is than attached to a microsphere and the microsphere suspended in a flow stream where it could be sequentially cleaved with an exonuclease. The cleaved fluorescent bases from the molecule would then be identified by their color as they pass through the laser beam, and the sequence in which the colors appear will describe the base sequence in the DNA. It is projected that between 100 and 1,000 bases per second could be sequenced by this technique.

## CELL DEATH

Having dealt with some of the more sophisticated aspects of DNA staining, I will end this chapter with one of the simpler (but often neglected) applications. Because propidium iodide is excluded from entering cells by an intact plasma membrane and because it only fluoresces when intercalated between the bases of double stranded nucleic acid, it will not fluoresce if it is added to a suspension of intact cells. The intact plasma membrane forms a barrier, keeping propidium iodide and nucleic acids apart. It is only when the outer membrane has been perforated that the cells will give off red fluorescence. Propidium iodide is therefore a stain (like trypan blue) that can be used to mark dead cells (on the reasonable but not necessarily valid assumption that cells with holes in their membranes large enough to allow the penetration of propidium iodide are actually dead according to other viability criteria and vice versa). By this method, cell death can be monitored in the presence of various cytotoxic conditions.

The only difficulty in using flow cytometry to monitor cell death is that, as mentioned in Chapter 3, dead cells have different scatter properties from living cells. In particular, because of their perforated outer membrane, they have a lower refractive index than living cells and therefore have an FSC signal of lower intensity. For this reason, it is important not to use a gate or FSC threshold when analyzing a population for the proportion of dead and live cells. Any FSC vs. SSC gate drawn around normal lymphocytes, for example, will always show most if not all of the cells within that gate to have excluded propidium iodide no matter how many cells in the preparation are dead; this is simply because the dead cells drop out of the gate (Fig. 7.21). The lesson here (and one that needs repeating in reference to most flow analysis) is that gating is an analytic procedure that needs to be performed carefully and with explicit purpose; if we do not define our population of interest (by gating) with care, our results may be accurate answers to inappropriate questions (*e.g.*, "what percent of living lymphocytes are alive?").

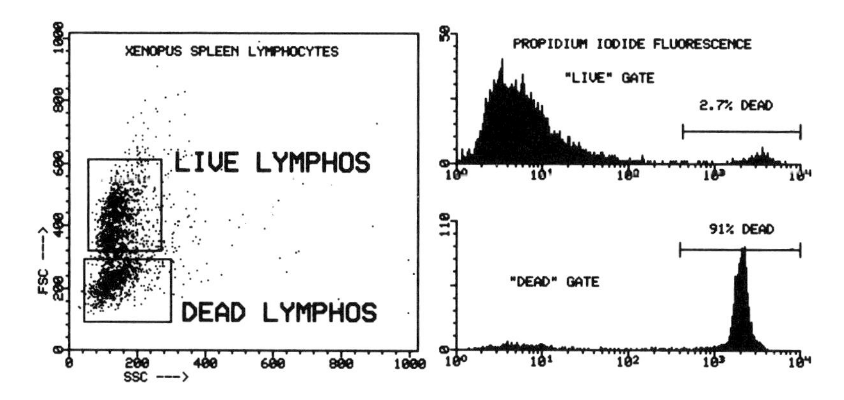


**Fig. 7.21.** The use of PI to monitor cell death. Dead lymphocytes have a less intense FSC signal than do live lymphocytes. Cultured *Xenopus* spleen lymphocytes courtesy of Jocelyn Ho and John Horton.

The use of propidium iodide to mark dead cells has a more routine application in simply allowing the exclusion of dead cells from analysis. This is important because dead cells have the unfortunate habit of staining nonspecifically when their broken membranes tend to trap monoclonal antibodies directed against surface antigens. Therefore a cell suspension including many dead cells may show high levels of nonspecific stain. By adding propidium iodide to cells just before analysis, the cells fluorescing red (>605 nm) can be excluded from further analysis by gating, and only the living cells then examined for surface marker staining. Figure 7.22 shows the way that this technique can be used when looking at FITC-staining for surface antigens on cultured mouse thymocytes. The use of propidium iodide to exclude dead cells from analysis is a procedure that is recommended as routine for any system using FITC staining of surface markers. It is particularly important in analysis of mixed populations where a high percentage of the cells may be dead. It can also be used in a three-color system, where two photodetectors register FITC and PE signals, and the third is reserved for the propidium iodide signal (the PE tube will also register a great deal of the propidium iodide signal, but as long as all propidium iodide-positive cells are gated out of further analysis, this will not lead to any problems). It should, however, be remembered that fixed cells cannot be stained with propidium iodide for live/dead

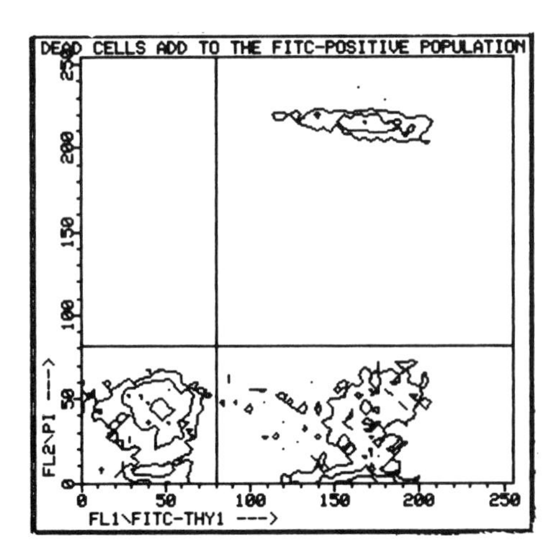


**Fig. 7.22.** The use of PI to exclude dead cells from analysis of a mouse spleen cell population for the expression of the Thy-1 surface antigen. The cells were stained with FITC for Thy-1 and then with PI to mark the dead cells. It can be seen that the dead cells (upper right quadrant) contribute to the FITC-positive population. Stained cells courtesy of Maxwell Holscher.

discrimination. Fixed cells are, in fact, all dead, and will therefore all take up propidium iodide even if some were alive and some dead before fixation.

## FURTHER READING

Chapters 13–16 and 24 in **Melamed et al.** are good reviews of the applications of flow cytometry to various aspects of cell cycle research.

Molecular biology and chromosome analysis are discussed in Chapters 25–28 of **Melamed et al.** A special issue (Volume 11, Number 1) of the journal **Cytometry** is devoted to the subject of analytical cytogenetics; it contains articles about work at the interface between advanced theory and clinical practice (as well as some beautiful pictures).

Good discussions of the mathematical algorithms for cell cycle analysis can be found in Chapter 6 of **Van Dilla et al.,** Chapter 23 of **Melamed et al.,** and in a multiauthor book called ***Techniques in Cell Cycle Analysis*** (1987), by JW Gray and Z Darzynkiewicz (eds), The Humana Press, New Jersey.

Slit-scanning is described in more detail in Chapter 5 of **Van Dilla et al.** and in Chapter 6 of **Melamed et al.**

Chapter 4 in **Ormerod** and several sections in the ***Purdue Handbook*** give good practical protocols. **Darzynkiewicz and Crissman** include many chapters with protocols on both the simpler and the more complex methods in nucleic acid and cell cycle analysis.

The proposed technique for sequencing DNA is described in a paper by Jett JH, Keller RA, Martin JC, et al. (1989). High-speed DNA sequencing: An approach based upon fluorescence detection of single molecules. J. Biomol. Struct. Dynamics 7:301-309.

# 8

# Disease and Diagnosis: The Clinical Laboratory

Early in this book it was stated that one of the areas in which cytometry was beginning to have great impact is in clinical diagnosis. Research work, beginning with Kamentsky's but particularly work during the last decade, has made it clear that the information that can be obtained by flow cytometry could be of obvious use to clinicians. But as long as flow cytometers maintained their image of being cumbersome and fiddly instruments, difficult to standardize and requiring constant attention from a team of unconventional but devoted hackers (see Figures 1.3 and 1.4 for an understanding of the origins of this image), there was little chance that cytometry would become integrated into a routine hospital laboratory. The direct cause of the recent rapid introduction of flow technology into many hospitals has been the commercial development and marketing of "black box" cytometers—instruments whose installation simply involves placing them on a laboratory bench and plugging them in; instruments designed to be stable and that therefore can, in principle, be maintained with relatively little human intervention; instruments that will produce clinical printouts without any requirement for knowledge of the intricacies of flow data analysis; and instruments that, we are told, can be run by anyone with the ability to push a key on a computer keyboard (Fig. 8.1). It was only with the development and marketing of these friendly machines that provision of flow cytometric information for routine diagnosis became a practical proposition.

As a result of the introduction of cytometers into the hospital setting, three aspects of clinical practice have led to some general reassessment of the nature of flow analysis. First, clinical laboratories are, because of the import of their results, overwhelmingly concerned with so-called quality control. This concern has forced all cytometrists to become more aware of

**Fig. 8.1.** Two opposing fantasies of what flow cytometry is all about. Drawings by Ben Givan.

the standardization and calibration of their instruments. Neither standardization nor calibration comes naturally to a flow cytometer. In response to this clinical requirement, beads, fixed cells, and mock cells have been developed to help in assessing the constancy of conditions from day to day. Instruments can now be set up from stored computer information so that machine parameters are constant from run to run. Furthermore the analysis of data can be automated so that clinical information can be derived and reported quickly and in standard format. In addition, quality control schemes of various sorts have resulted in samples flying around the world; reports are appearing in the literature documenting the factors that lead to variation in results obtained from different laboratories and with different operators. Quality control in flow cytometry is still in its infancy, but progress is being made toward producing technical recommendations (both for DNA analysis and for lymphocyte surface marker analysis), toward providing a scheme for accrediting personnel, and toward monitoring the performance of laboratories—both between laboratories and within any one laboratory over the course of time. Quality control schemes have been organized in the United States by a committee of the International Society for Analytical Cytology in cooperation with the National Committee for Clinical Laboratory Standards.

The second aspect of clinical practice that has led to a reassessment of the nature of flow cytometry is the large amount of data acquired in routine use and the need for a way to store these data over a long period of time, correlate them with various factors, and maintain some sort of running total—in short, the need for a flow database. This clinical requirement has led to the development of programs for integrating flow data into established software databases. It has also led to the provision of programs for sending such data into personal computers for analysis away from the cytometer itself. Needless to say, such developments are as useful in a research environment as in the hospital community.

The third aspect of clinical practice that has led to modifications in flow technology has been the requirement for safety in the handling of potentially infected specimens. The fluidic specifications of instruments have been modified with attention to the control of aerosols that might occur around the stream, the prevention of leaks around the sample manifold, and the collection of the waste fluid after it has been analyzed. But a more effective means of minimizing biological hazards has been the development of techniques for killing and fixing specimens in such a way that cells, viruses, and bacteria are no longer viable but the scatter and fluorescent properties of the cells of interest are relatively unchanged. Paraformaldehyde (0.5%–1.0%) is the fixative of choice in most flow laboratories. It is used after cells have been stained. As well as destroying the infectivity of the hepatitis B and AIDS (HIV) viruses, it fixes lymphocytes in such a way that their scatter characteristics are virtually unchanged. Moreover, the fluorescence intensity of their surface stain remains essentially stable (albeit with slightly raised control autofluorescence) for several days or more. Therefore routine fixation of biological specimens not only has increased the safety of flow procedures but also has made it possible to ship specimens around the world and to store specimens within a lab for convenient structuring of access to the cytometer (that is a euphemism for not working in the middle of the night). It should be said, however, that anyone working with material of known biological hazard needs to check fixation procedures to confirm the loss of infectivity. Even after fixation, anyone working with any biological material at all should use standard precautions for control, since any particular fixation procedure might not be effective against unknown or undocumented hazards. And whatever fixation procedure is used should be checked with individual cell preparations and staining protocols to confirm the stability of cytometric parameters over the required period of time.

Hospital cytometers have been appearing in three general types of laboratory. I will describe each of these briefly in turn.

## THE HEMATOLOGY/IMMUNOLOGY LABORATORY

The flow cytometer has, for several reasons, begun to find a very natural home in the hospital hematology laboratory. In the first place, because of the virtually universal use of automated blood analysis instrumentation for enumerating erythrocytes and leukocytes, people working in hematology laboratories are quite relaxed about the idea of cells flowing in one end of an instrument and numbers coming out the other end. As I have said, both historically and technologically there is a close relationship between flow cytometers and Coulter counters. The second reason for the hematologist's ease with flow technology is that hematology/immunology laboratories were among the first to make routine use of fluorescently tagged monoclonal antibodies. For many years, hematologists have been using panels of monoclonal antibodies for identifying various subpopulations of lymphocytes that are suggestive or diagnostic of various disease conditions. In particular, since the leukemias and lymphomas represent a group of diseases that involve the uncontrolled clonal proliferation of particular groups of leukocytes, various forms of the leukemias and lymphomas can be identified and classified according to the phenotype of the increased number of white cells found in biopsy material or in the patient's peripheral circulation.

Studies of the staining of surface markers to determine the phenotypes of various leukemias/lymphomas have been usefully circular, revealing much about ways to classify the diseases but also increasing to a considerable extent our knowledge of normal immune cell development. In general, immune cells gain and lose various surface proteins in the course of their normal development in the bone marrow until they become the cells with mature phenotype and function that are released from the marrow into the peripheral circulation (see Fig. 8.2). Leukemias and lymphomas often involve a block in this development so that cells with immature phenotypes appear in great numbers in the periphery. Alternatively, certain forms of disease may involve rapid expansion of a clone of mature normal cells so that one type of cell predominates in the circulation over all other normal subpopulations. In other cases, cells seem to express abnormal combinations of markers, a condition referred to as *lineage infidelity* or as *lineage promiscuity* depending on one's interpretation of the data. This makes classification of leukemias and lymphomas according to their surface membrane markers a suitable application for flow cytometry, albeit a very complex task. Correlation between classification and clinical course is therefore correspondingly difficult. Flow cytometric techniques for analysis of DNA content and BrdU incorporation have recently begun to be used, in conjunction with surface marker analysis, to help with this challenge.

The other condition that involves analysis of peripheral blood leukocytes is AIDS. Early in the natural history of the disease (or at least in the natural history of immunologists' awareness of the disease) it was discovered that one subpopulation of T lymphocytes in particular was destroyed by the HIV virus; the cells destroyed are those that possess the CD4 protein on their surface. It is this CD4 protein that appears to be the receptor involved in virus targeting. Therefore much of the diagnosis and staging of AIDS involves the enumeration of $CD4^+$ cells in the peripheral blood (Fig. 8.3).

All of these techniques—for counting $CD4^+$ cells in connection with AIDS diagnosis and for counting various populations of leukocytes for leukemia/lymphoma diagnosis—were performed routinely in hematology laboratories by the staining of cells with fluorochrome-conjugated monoclonal antibodies followed by the visual identification of different types of white cells and the counting of the fluorescent *vs.* unstained cells under the microscope. While the microscope has certain very definite

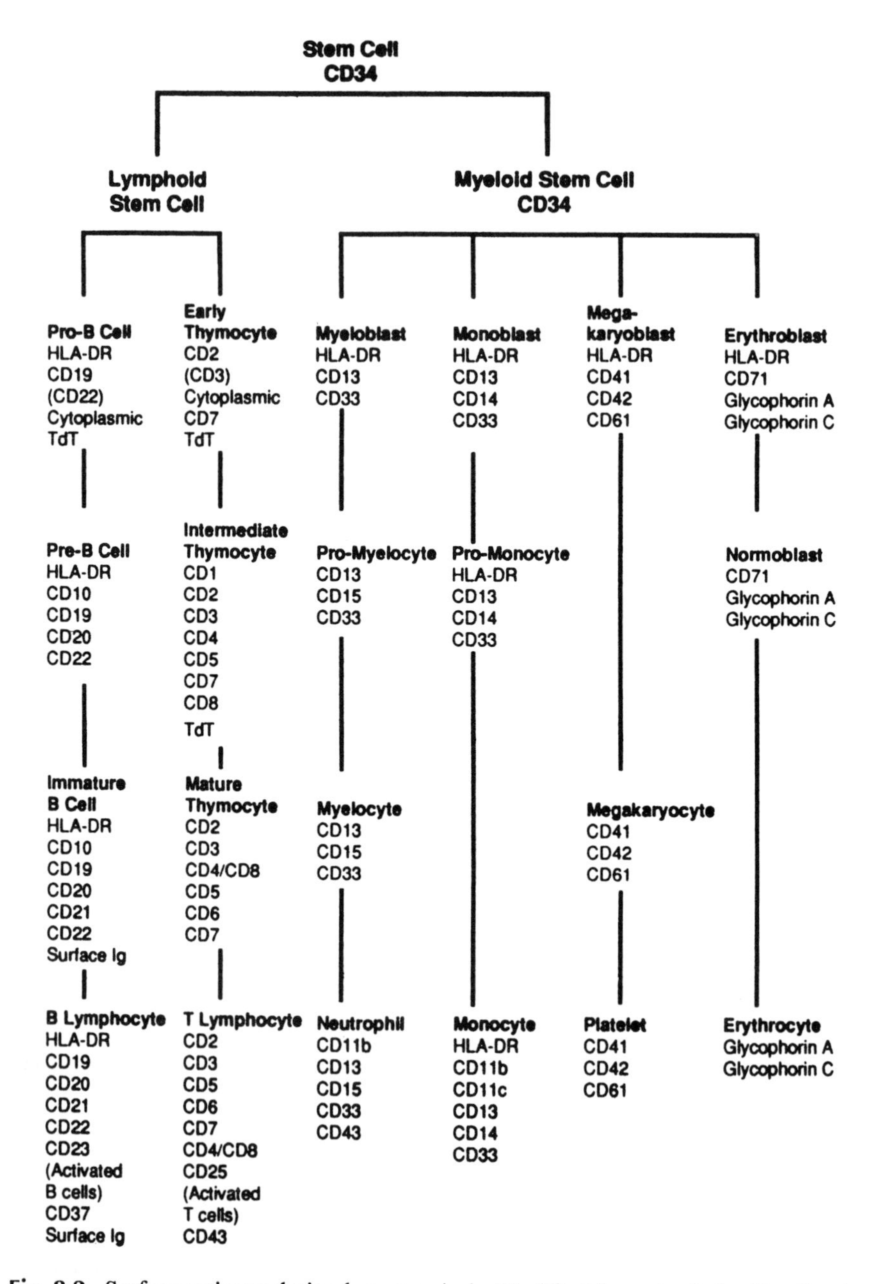


**Fig. 8.2.** Surface antigens during hematopoiesis. Modified from the Dako *Handbook on Flow Cytometry.*

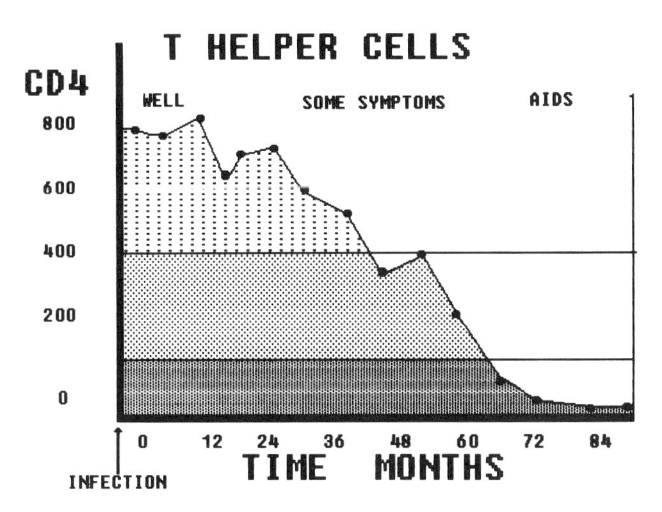


**Fig. 8.3.** The number of T-helper ($CD4^+$) lymphocytes ($\times 10^3/cm^3$) in peripheral blood of a patient as months after infection with HIV. Cells <400 are significantly below the normal range; <100 indicates severe risk of clinical AIDS. Figure courtesy of Léonie Walker.

advantages over the flow cytometer, two advantages it does not have are those of speed and statistical reliability. Particularly as a result of their work load from the growing number of AIDS patients, the hematologists' need for a way to count statistically reliable numbers of cells from large numbers of patients has become increasingly urgent. Flow cytometry is the obvious answer to this need. At the present time, more and more hematology/immunology laboratories are acquiring flow cytometers.

Although the initial perceived need was for the rapid processing of samples from leukemic and HIV-positive patients, as is the way with flow cytometry, the presence of the cytometer has stimulated thought about new hematological applications. For example, the staining of red cells for RNA content with a dye called thiazole orange has made possible the use of flow cytometry to count the reticulocytes present in blood samples from anemic patients. Flow cytometry is also used to confirm the depletion of classes of lymphocytes before bone marrow transplantation. Hematologists have extended the use of flow analysis to platelets—those particles with low FSC that usually are ignored in flow cytometric applications because they fall below the FSC threshold. Platelet-associated immunoglobulin and antiplatelet antibodies can be measured by flow cytometry. The former (indicative of autoimmune problems) is determined by using FITC-conjugated antibodies to detect immunoglobulin on the surface of the patient's platelets. The antibodies in the serum with specificities for "foreign" platelets that may develop after pregnancy or transfusions can be monitored by flow cytometry if a patient's serum is incubated with a donor's platelets.

## ONCOLOGY/PATHOLOGY LABORATORIES

In the chapter on DNA analysis, I mentioned the large amount of work generated for flow laboratories as a result of publication of the Hedley technique for analyzing the DNA content of paraffin-embedded pathology specimens. After the first headlong rush of publications correlating DNA ploidy with long-term prognosis in various types of cancer, the field has now settled down a bit. As I have indicated, it would probably be fair to say that most (but not all) studies on disaggregated solid tumors (fresh, frozen, or fixed) have shown some kind of correlation between abnormal DNA content and unfavorable long-term prognosis. There is considerable debate in the literature about whether flow cytometric analysis of ploidy gives any additional prognostic information that is independent of other known prognostic indicators (such as lymph node status in breast cancer), but most studies have indicated that it does. It appears to be useful, for example, in indicating, from among all breast cancer patients without lymph node involvement, a group of women who are at risk for recurrence of disease (Fig. 8.4). Current clinical research now centers around three aspects of DNA analysis. First, because any pathological material will be a mixture of both normal and abnormal cells, there is much effort expended on ways of selecting from among the cells of a disaggregated specimen those particular cells that are from the tumor itself. If those tumor cells can be stained selectively (as with a FITC-conjugated anti-cytokeratin antibody for epithelial cell–derived tumors within nonepithelial tissue) and then the DNA content of the FITC-stained cells determined, we may have a dual color flow method of much improved sensitivity for detecting aneuploid cells from within a mixture of both malignant and normal components.

The second aspect of pathology that has received much attention is that of quality control. Quality control presents a formidable challenge in DNA analysis for several reasons. For one thing, the quality of fixed, embedded specimens is variable and the propidium iodide (PI) fluorescence intensity of fixed material erratic; thus obtaining an appropriate control of normal tissue

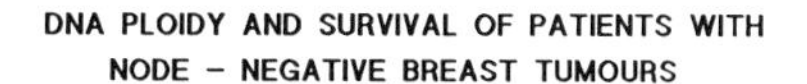


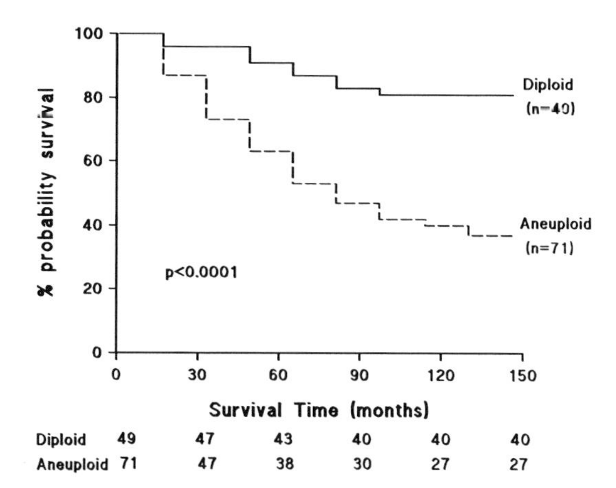


**Fig. 8.4.** Survival curves for breast cancer patients without lymph node involvement. Those with aneuploid tumors (as diagnosed by PI flow cytometry of 10-year-old paraffin-embedded material) survived significantly less long than those with diploid tumors. Graph from Yuan et al. (1991).

is not usually possible. For this reason, classification of aneuploid specimens from fixed material is, for the most part, pragmatic and depends on the presence of two peaks in the DNA histogram. But material of poor quality usually produces peaks with wide CVs, and such peaks make it difficult to detect near-diploid aneuploidy. The debate about quality control in DNA analysis and how to standardize classification is one that will, I am sure, run and run. Some of the heat in this debate, however, merely reflects the ease of determining ploidy by flow cytometry and the general enthusiasm within the medical community for rapid implementation of the technique.

Finally there is evidence that determination of the proliferative potential of malignant cells may be a better indicator of poor prognosis than simply the determination of the presence or absence of aneuploid cells. As discussed in the chapter on DNA, proliferation can be assessed by looking at the shape of the DNA histogram. However, the presence of more than one cell line, with the G2/M peak from the diploid cell line overlapping the S phase of the aneuploid cells, makes mathematical analysis even less reliable than with distributions resulting from single euploid cells or clones. BrdU has been used clinically to give more idea about proliferation; it has been used both *in vitro* with fresh excised tumors and *in vivo* by infusion into patients at some time prior to excision of the tumor. Estimates of tumor doubling time, based on the rate of movement of BrdU-pulsed cells through the cell cycle and on the percent of cells in the tumor that are dividing, have been shown to correlate with aggressiveness of the malignancy. There may also be applications of this technique for assaying malignancies for sensitivity to various cytotoxic drugs. In this field, recent research has developed the use of monochlorobimane (which fluoresces in the presence of glutathione) to look at glutathione levels in cells as an indicator of drug resistance.

Most of these techniques still come under the heading of clinical research and are not yet in the category of routine hospital service. Ploidy determinations are, in most cases, the only flow analyses that have achieved much routine application in the field of oncology. At the present time, it would appear that flow cytometry has not settled quite so comfortably into the pathology laboratory as it has into the hematology/immunology setting. This may be partly a result of the quality control problems that come from using archival specimens. It may also, I suspect, come from the crucial difference between hematology and pathology—and it is a difference that may remind us of one very definite limitation of flow analysis. Pathological specimens come from solid tissue; to analyze them with a flow cytometer, the solid tissue must be disaggregated in some way. This disaggregation process (either mechanically, with detergent, or with enzymes) not only has a good chance of compromising the integrity of some or many of the cells we want to analyze but also will, necessarily, destroy any information that is contained in the structural orientation or pattern of the original tissue. A great deal of information can be derived from these patterns. One strength of pathologists is their ability to draw conclusions from looking at tissue sections. Flow cytometry will never be as useful in analyzing solid tissues as it is at analyzing cells that come naturally in single cell suspensions.

## TRANSPLANTATION

It has been known for many years that transplant recipients may possess serum antibodies that will react with and destroy cells from a transplanted organ. If these so-called cytotoxic antibodies are present at the time of transplantation, then an immediate and violent rejection crisis will occur (hyperacute rejection), destroying the grafted organ and threatening the life of the recipient. Therefore, the serum from a potential recipient is assayed before surgery for cytotoxic antibodies by mixing it with lymphocytes from the organ donor (in the presence of rabbit complement). Traditionally, the cytotoxicity of the serum is scored by counting dead lymphocytes (cells taking up PI or trypan blue) under the microscope.

In the early 1980s, Garovoy and his group in California suggested that this cross-match assay might not be sensitive enough. Some of the cases of slower rejection might conceivably result from the presence in the recipient of levels of preformed antibodies that were too low to be detected by the traditional cytotoxic assay. A flow cytometric cross-match assay for these preformed antibodies directed against donor lymphocytes was developed by Garovoy and subsequently modified by Talbot in Newcastle and by workers at other centers to make use of two-color fluorescence. The assay involves the incubation of the organ donor's lymphocytes with serum from potential recipients followed by the staining of the lymphocytes with a phycoerythrin-conjugated monoclonal antibody against either T cells or B cells and an FITC-conjugated antibody against human immunoglobulin (Fig. 8.5). In that way, it can be determined whether the recipient has

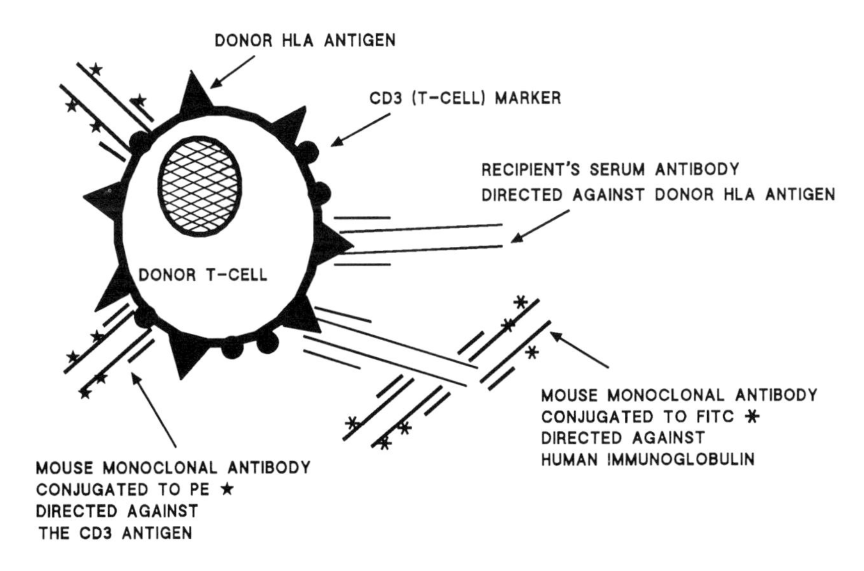


**Fig. 8.5.** Protocol for the dual-color flow cytometric assay for testing recipient serum for the presence of antibodies to donor cells before organ transplantation.

antibodies that coat the T lymphocytes or the B lymphocytes of the donor (Fig. 8.6). This two-color technique can in fact serve to remind us of a general approach to the design of flow cytometric assays: One color is used to pinpoint the cells of interest (in this case, either T cells or B cells), and the second color is used to ask some question about those cells (in the cross-match; we are asking whether those T or B cells have become coated with immunoglobulin from the recipient's serum).

Evidence over the past 5 years has indicated that this assay for lymphocytes with bound antibody is indeed more sensitive in detecting antibodies directed against donor cells than is the traditional assay for lymphocyte death. And the flow assay does define a group of recipient–donor pairs who are at risk for severe rejection crises and possibly, but not necessarily, for loss of the grafted organ. At the present time, donated organs are in short supply. One transplant surgeon has stated persuasively that, in selecting patients for transplantation, he is most concerned with making sure that the donated kidney will survive. Although the flow cytometric cross-match assay may arguably be denying some recipients unnecessarily the opportunity of a successful transplant because it may be too sensitive, it certainly is helping surgeons to find a group of patients who have the very best chance of accepting the transplanted organ. Because transplant surgeons have few organs to graft and long waiting lists of potential recipients, and because not enough is yet known about ways to predict, diagnose, and

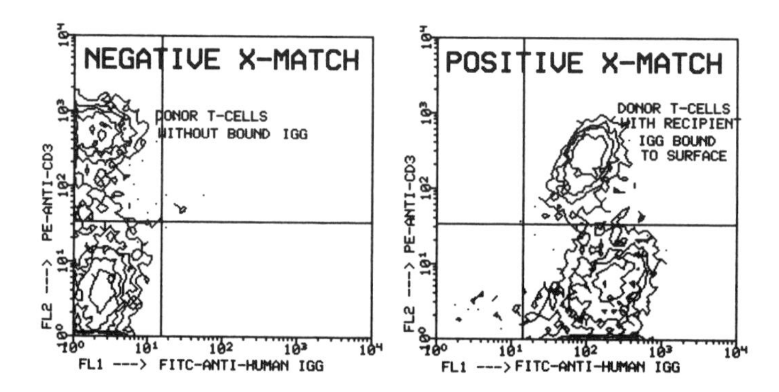


**Fig. 8.6.** Contour plots indicating a cross-match assay for immunoglobulin from the recipient bound to the surface of the donor's T cells. Data from the Tissue-Typing Laboratory, Royal Victoria Infirmary, Newcastle upon Tyne.

abrogate rejection, the use of flow cytometry has been accepted relatively quickly into routine use by the transplant community.

The acceptance of the flow cross-match into the hospital environment highlights some of the difficulties that occur when any technique moves from research into service use; in the case of transplantation surgery, some of these difficulties are particularly acute. The flow cross-match needs to be implemented in such a way that it can be performed by people who are not flow cytometry specialists and who may be working in the middle of the night under pressure of time and without specialist support staff available for easy consultation. Under such conditions, automated software and stable instrumentation are especially important. In addition, flow cytometers are relatively expensive instruments; cross-match laboratories are concerned with ways to justify the expense of a cytometer because it needs to be available all the time, but it may only be used on infrequent occasions. However, the most serious problem in implementing the flow cross-match for routine use has been in defining the position of the dividing line between negative and positive results. With increasing instrument sensitivity, large numbers of patients will show some level of serum antibodies directed against donor cells. Definition of a fluorescence intensity borderline for proceeding with or denying an organ transplant clearly has to involve some understanding of the clinical (and ethical) requirements of patients and of the supply and demand for organs as well as of the technological capabilities of the cytometer.

## COMMENTS

Rational decisions still need to be made about the fields in which flow cytometry can make a positive contribution to patient care, about the methods available for quality control, and about the kind and depth of training required by staff. Good channels of communication also need to be opened between laboratory staff and the clinicians who are making diagnostic use of flow data, but who may not be aware of its limitations and conventions. At the beginning of this chapter were described the new class of clinical cytometers that are reasonably stable and can be used and maintained with little human intervention. Software packages for those cytometers can apply, *e.g.*, algorithms for automated lymphocyte gating of mononuclear cell preparations—the computer will draw a lymphocyte gate for you. The future will contain more sophisticated expert systems employing artificial intelligence in order to avoid the pitfalls associated with human judgement. How stable these instruments actually become, whether automated computer algorithms make reasonable guesses with difficult samples, and whether an untrained scientist or clinician can draw reliable conclusions from automated print-outs of flow data are open questions. We all need to guard against that demon of a false sense of security.

Ken Ault, the former president of the International Society for Analytical Cytology, summed up some of flow cytometry's clinical growing pains with the following statement:

> Flow cytometry is a technology that seems to stand at the threshold of "clinical relevance." Those of us who have been using this technology, and especially those who are manufacturing and selling the instruments and reagents, are frequently evaluating the status of clinical flow cytometry. Most of us have little doubt that this is going to be an important technology in clinical medicine for many years to come. However, from my point of view, and I think for many others, the movement into the clinic has been unexpectedly slow and painful....In talking about this in the past I have frequently mentioned the "grey area" between research and clinical practice. I was recently reminded of a quotation from T.S. Eliot: "Between the idea and the reality falls the shadow." I believe that flow cytometry is currently traversing that shadow. Ault KA (1988). Cytometry (Suppl 3):2–6.

That paragraph was written for a talk given in 1986. Clinical cytometry is now, in 1992, in what we might call dappled sunlight; however, the full glare of total integration into the hospital community is still in the future.

## FURTHER READING

Supplement 3 of the journal *Cytometry* (1988) and Volume 6 (Number 1) of *Seminars in Diagnostic Pathology* (1989) are both devoted to clinical applications of flow cytometry.

Chapters 35–39 in **Melamed et al.** discuss various aspects of clinical oncology, as does chapter 15 in **Watson**.

An article by David Hedley (1989) in *Cytometry* 10, pages 229–241, entitled "Flow cytometry using paraffin-embedded tissue: five years on," gives a good review of this clinical field.

A book entitled *Flow Cytometry* in the series "Guides to Clinical Aspiration Biopsy" by Philippe Vielh (Igaku-Shoin, Tokyo, 1991) has good discussions of clinical problems of DNA analysis and some very nice DNA histograms paired with micrographs of the stained cells.

The original article on the use of flow cytometry for the transplantation crossmatch is Flow cytometry analysis: A high technology crossmatch technique facilitating transplantation. Garovoy MR, Rheinschmidt MA, Bigos M, et al. (1983). Transplant. Proc. 15:1939. The two-color assay is described in Talbot D, Givan AL, Shenton BK, et al. (1988). Rapid detection of low levels of donor specific IgG by flow cytometry with single and dual colour fluorescence in renal transplantation. J. Immunol. Methods 112:279–283. A more recent article with some discussion of possible ways of classifying the flow data is Mahoney RJ, Ault KA, Given SR, et al. (1990). The flow cytometric crossmatch and early renal transplant loss. Transplantation 49:527–535.

# 9

# Out of the Mainstream: Research Frontiers

In the previous chapters, I have discussed what may be thought of as the mainstream applications of flow technology. Cells stained for surface antigens and nuclei stained for DNA content together constitute a large majority of the particles that flow through the world's cytometers. As noted in the Preface, however, flow cytometry has continued to surprise everyone with its utility in unusual and unpredicted fields of endeavor. By the time this book appears in print, some new applications will almost certainly have progressed into the flow mainstream and other newer applications will have taken their place on the fringe. At present, the large number of "fringe" applications (highly varied, often idiosyncratic, and sometimes unsung) precludes any attempt at making a chapter in an introductory book such as this into a complete catalogue. Therefore, I will attempt here only to hint at the wide range of uses that exist by providing a taste of a few of the applications that seem to me to be important, or unusual, or to show promise. Because each reader will have his or her own personal goals for the use of flow cytometry according to his or her own interests, I think it is important here mainly to convey some feeling for the enormous variety and surprising potential that are found in laboratories with an interest in flow research.

## FUNCTIONAL ASSAYS

Whereas the applications discussed earlier in this book have dealt with ways to describe cells or nuclei that have been stopped in their tracks for analysis at a given moment, flow cytometry has also been used to follow the physiological function of cells in kinetic analysis. As examples of this kind of analysis, I will discuss the use of a cytometer to look at the changes in calcium concentration that occur when cells become physiologically activated.

One of the early responses made by many types of cells to a stimulus is a rapid increase in the influx of calcium ions across the plasma membrane and a resulting increase in the level of free cytoplasmic calcium. This increase is thought to be one of the steps involved in so-called signal transduction and can result in the activation of enzyme systems responsible for subsequent metabolic or developmental changes. Lymphocytes show increases in intracellular calcium in response to many kinds of specific and nonspecific surface ligand binding, some of which lead to the cellular changes that we associate with an immune response.

Many classes of cells, including neutrophils, nerve cells, and platelets, also show calcium changes in response to stimulation. A range of dyes developed by Roger Tsien in California has been useful in flow cytometry because they can be loaded into living cells where they will chelate calcium in a reversible equilibrium and fluoresce in proportion to their calcium load. A dye called *fluo-3* absorbs light from the 488 nm line of the argon laser and fluoresces little in the absence of calcium but significantly (at 525 nm) when binding the ion (Fig. 9.1). However, the use of fluo-3 is difficult to standardize because the amount of fluorescence varies with the amount of dye loaded into the cells. A

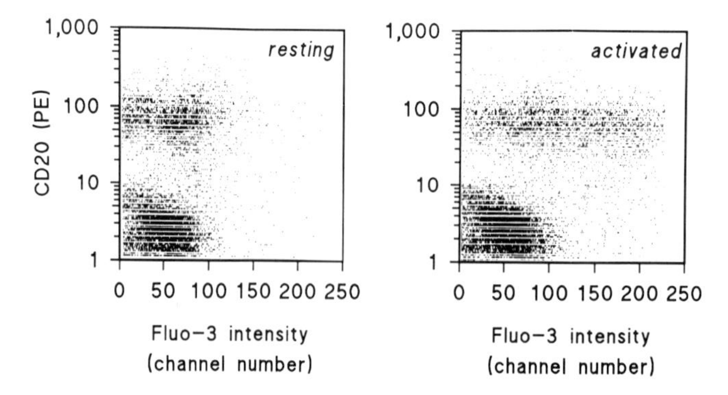


**Fig. 9.1.** The calcium level (as measured by fluo-3 fluorescence) in peripheral blood B cells (CD20-PE$^+$) before and after addition of an anti-μ antibody. Figure courtesy of GT Rijkers.

more useful dye is the related indo-1, which, although it requires a light source with UV output for excitation (a high-power argon laser or a mercury arc lamp, for example), permits a so-called ratio analysis of calcium. The term *ratio* in this context simply means that indo-1 fluoresces at 480 nm (turquoise) when free of calcium but at 405 nm (violet) in the chelated state. By using the ratio between violet and turquoise fluorescence, we can get a measure of the amount of calcium in a cell that is independent of the amount of dye loaded. Some flow cytometers can calculate the ratio between any two of their parameters. For others, this calculation has to be done manually. Figure 9.2 shows a time course of this ratio (proportional to internal calcium) as it changes in response to the stimulation of lymphocytes by a ligand bound to their surface receptors. Such a time course can be performed on a flow cytometer by loading cells with indo-1 in advance and then stimulating them with a ligand (in a calcium-containing medium) immediately before placing the sample tube on the sample manifold. Alternatively, the sample manifold of many cytometers can be modified to permit rapid injection of reagents into the sample tube while the sample is being drawn through the system. In either case, measured cell parameters can then be acquired over the subsequent minutes and their 405:480 fluorescence ratio determined.

Of course, in order to talk about kinetic measurements, we need to bring in the parameter of time. Time is, in many ways, the hidden extra parameter

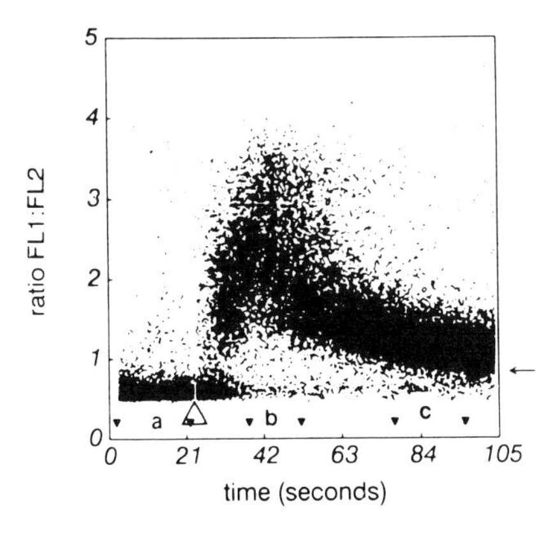


**Fig. 9.2.** The time course of a calcium response (measured by the indo-1 violet/turquoise fluorescence ratio) induced in a B-cell line by addition of anti-μ antibodies. From Rijkers et al. (1989).

of every flow system. Computers have in-built clocks, and for every cell acquired into the computer memory there will be, along with the four or more light-based parameters that we have talked about, a time of acquisition attached to it in its data file. Depending on the software being used, it will be more or less easy to access this time parameter for use in a kinetic profile. In the best case, we will be able to plot any other parameter(s) like turquoise and/or violet fluorescence against time for each cell acquired into the computer memory, thus giving a time course of the change of that parameter during the course of the sample acquisition. In a system with less sophisticated software, we may simply be able to acquire and store in data files repeated samples of, say, 2,000 cells and then, during analysis, determine both the median fluorescence of those cells and the beginning and end time of the entire sample acquisition. In that way, we could plot fluorescence (or ratio of fluorescences) against the start time for each sample and would be able to analyze the rate of reactions—provided only that the rate we are considering is relatively slow compared with the time that it takes to acquire each sample of 2,000 cells (Fig. 9.3).

This kind of procedure is an example of a class of protocols that measure functional activity in living cells. They depend on the ability to find a compound able to enter living cells that will then alter its fluorescence properties in relation to a changing intracellular environment. As we have seen, compounds are available that will serve to assay calcium concentration. Other compounds will respond to pH or to changes in membrane potential in similar ways. The uptake of fluorescent beads has been used to measure the phagocytic activity of cells. Reduced fluorochromes such as dichlorofluorescin diacetate, which can be loaded into cells and which fluoresces only when oxidized, have been used to measure the production of peroxides during neutrophil activation. And Herzenberg's group has developed techniques for looking at cells for the presence of enzymes like β-galactosidase (and, by extension, for expression of the transfected genes for these enzymes) with fluorogenic substrates. In general, the development of functional probes is a field in which organic chemists and flow cytometrists are working closely. New probes are being developed all the time to suit new functional applications, to exploit new excitation light sources, and to satisfy the increasing desire on the part of scientists for the power that results from the simultaneous analysis of both functional and phenotypic parameters. Almost everyone starts out in flow cytometry with work on static and possibly fixed or dead cells. A shift of direction into the realm of function and kinetic analysis

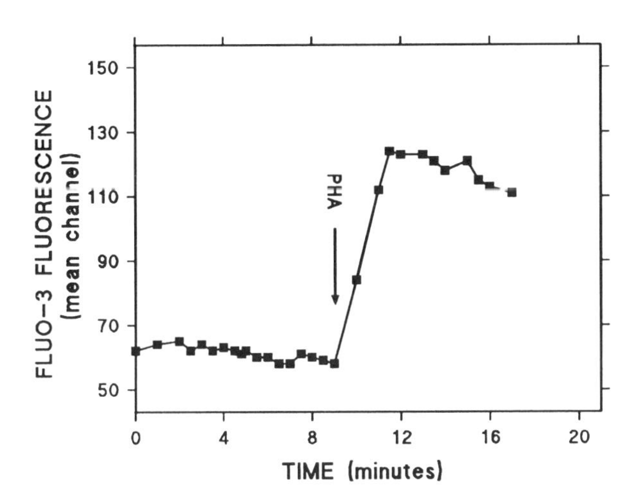


**Fig. 9.3.** A time course for the calcium change (measured by fluo-3 fluorescence) induced by the addition of phytohemaglutinin (PHA) to peripheral blood lymphocytes. Two thousand cell data files were acquired repeatedly from a continuously running sample. The mean fluorescence for each group of 2,000 cells was then plotted against the time of the start of acquisition of each data file.

requires a broadening of one's entrenched ideas of what flow cytometry is all about. The shift is almost certain to prove both challenging and informative.

## THE AQUATIC ENVIRONMENT

Blood has been an ideal object for flow cytometric attention in large part because it occurs naturally as a mixed population of single cells. Lakes and oceans are also suspensions of mixed single-celled particles. As such, they would appear to present an obvious target for flow analysis. Indeed, flow cytometers are now in place in many marine laboratories and on board sea-going vessels.

Aquatic single-celled organisms with a size range of 0.5–100 μm in diameter occur in nature in concentrations that range from about $10^2$ to $10^6$ per $cm^3$. They include bacteria, cyanobacteria (also known as *blue-green algae*), autotrophic phytoplankton (unicellular plants), and heterotrophic zooplankton (unicellular animals). The analysis of aquatic organisms by flow cytometry presents some characteristic features that may serve to highlight the issues that, to a greater or lesser extent, affect all flow analysis. In the first place, because of the presence of naturally occurring photosynthetic pigments, the phytoplankton are highly autofluorescent (some of them contain phycoerythrin). This autofluorescence leads to high background intensity against which positive staining of low intensity may be difficult to detect; the autofluorescence is also variable and may depend on the environment or metabolic state of the cell. However, the autofluorescence can be exploited and used to distinguish different classes of organisms and different metabolic states. Another characteristic of the aquatic environment is that the abundance of organisms of different types is highly variable; aquatic scientists do not have the benchmarks of a fairly tight "normal range" that clinical scientists depend on. In addition, the abundance of very small cells in the aquatic environment presents a challenge in instrument tuning and sensitivity. Not all cytometers can distinguish the forward scatter signal of nano- or picoplankton from optical noise or from particulate matter in the sheath stream, and, because of the great size heterogeneity of plankton, a cytometer for aquatic analysis must be able to cope with both small and large particles at the same time. A final problem is that the most common particles in aquatic samples are not living; they represent decaying organic matter, silica, or calcium-containing empty cell walls and suspended sediment, which are all difficult for a flow cytometer to distinguish from living cells.

Despite these problems, flow cytometry has had some noted successes in aquatic research, particularly in relation to studies on the phytoplankton. Because all phytoplankton possess chlorophyll, but only the cyanobacteria possess the phycobiliproteins, autofluorescence "signatures" from water samples, based on the chlorophyll (fluorescence >630 nm), phycoerythrin (fluorescence <590 nm), and FSC of particles, have been used to characterize the changes that occur in plankton at different depths or at different locations (Figs. 9.4 and 9.5).

Figure 9.6 shows an example of the way in which flow cytometric analysis can distinguish six different species of plankton in culture and define which of these species are favored by grazing marine scallops as a source of food. In fact, results such as these have been used to suggest modifications in the menu supplied to scallops being farmed in aquaculture tanks.

Flow cytometry has also led to the notable discovery, reported by Chisholm in 1988, of the existence of a novel group of small, prokaryotic

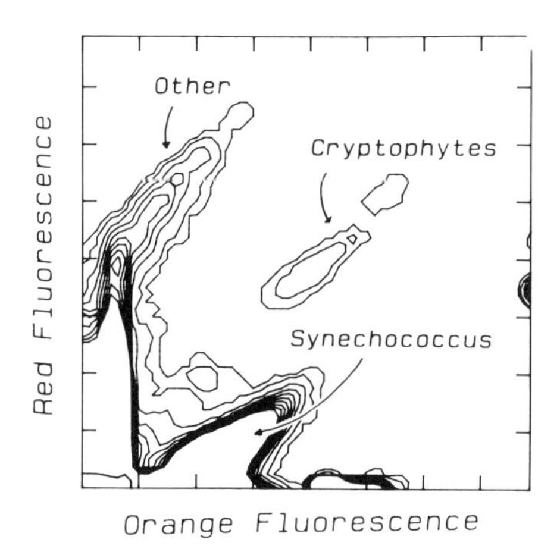


**Fig. 9.4.** The flow cytometric signature of a surface sea water sample from near shore off Woods Hole, MA. Cells with high chlorophyll content have intense red autofluorescence, and those with high phycoerythrin content have intense orange autofluorescence. From S Chisholm et al. (1986).

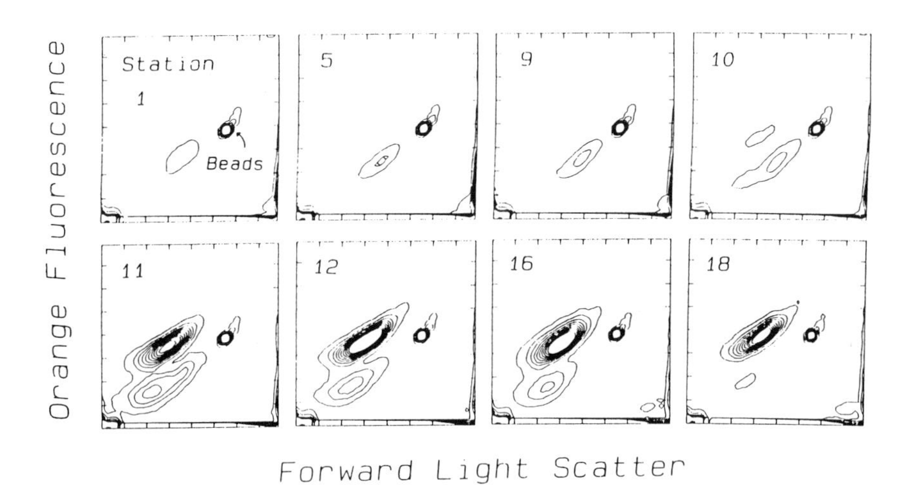


**Fig. 9.5.** Flow cytometric analysis of surface water from points at 1.5 mile intervals off shore from Cape Hattteras, NC. FSC and orange fluorescence identify two *Synechococcus* populations with different phycoerythrin content. Beads were used to calibrate the number of cells present. From Chisholm et al. (1986).

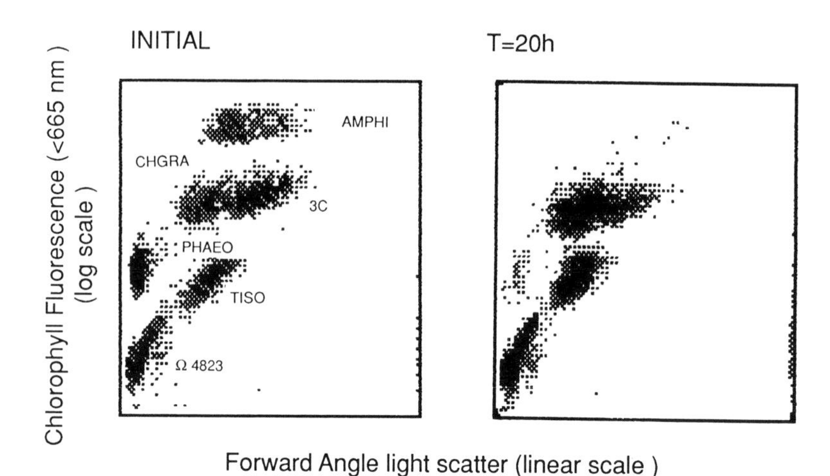


**Fig. 9.6.** Small scallops were placed in tanks with six species of phytoplankton that show distinctive flow cytometric signatures. After 20 hours of grazing, it was apparent from the differences between the flow dot plots that the scallops had exhibited definite preference for two of the six species. Figure courtesy of Sandra Shumway.

phytoplankton; these free-living, marine prochlorophytes are between 0.8 and 0.6 μm in size but possess pigments more like those of eukaryotic plants than of other prokaryotes. The use of shipboard flow cytometry during cruises off Southern California, the Panama Basin, the Gulf of Mexico, the Caribbean, and the North Atlantic between Woods Hole, MA and Dakar, Senegal (who said flow cytometry isn't fun?) has found these previously unknown prochlorophytes in remarkable abundance and indicated that they may be responsible for a significant portion of the global photosynthetic productivity of the deep ocean.

The problems presented by the heterogeneity of the aquatic environment and the instrumental and conceptual developments made by aquatic scientists toward handling these problems have led to advances that can enrich the work done in all fields of flow analysis. Cytometers that can deal with the instability of the shipboard environment will be all the more dependable in a relatively stationary land-based laboratory. Cytometers that are developed to handle both very small and very large particles may allow flow analysis to move more decisively into the fields of microbiology, parasitology, mycology, and botany. The general idea of studying autofluorescence instead of trying to avoid or ignore it is one that may be profitably considered by cytometrists in many areas of endeavor.

## ANIMAL DEVELOPMENT

In the section on functional assays, we mentioned the work of the Herzenbergs' group at Stanford in developing a system for assaying the presence of the enzyme β-galactosidase (coded by the *lac* Z gene from *Escherichia coli*). The presence of this enzyme can be detected flow cytometrically by use of a so-called fluorogenic substrate—in this case fluorescein digalactopyranoside (FDG), which is cleaved by β-galactosidase to fluorescein monogalactoside and then to fluorescein, which is fluorescent. The importance of assaying for the presence of β-galactosidase transcends any interest in regulation of expression of this enzyme in bacterial cells: The *lac* Z gene has been used extensively in molecular biology as a reporter for the presence and/or expression of recombinant genes in mammalian cells. Cloned mammalian genes can be inserted, along with the *lac* Z bacterial gene, into mammalian cells; if they are all under the control of the same promoter, expression of the *lac* Z gene will then become a marker for the expression of the cloned mammalian genes.

A creative use of this technique has been developed by Nolan and Krasnow at Stanford in a system they call *whole animal cell sorting* (WACS). The system does not involve sorting of intact animals (sheep to the left, goats to the right) but rather sorting of all the cells from a whole animal, after they have been dissociated. Specifically, the system has been used to study development of the fruit fly, *Drosophila*. Many of the identifiable cell types in developing *Drosophila* embryos have specific promoter regions in their genome that become activated in the course of development to initiate the formation of gene products typical of each cell type. Embryos can be transfected with *lac* Z as a reporter gene into chromosome positions driven by a cell-type-specific promoter. These embryos (containing the introduced *lac* Z gene under the control of a specific promoter) are then grown to a given developmental stage. The cells expressing the reporter gene will contain β-galactosidase. Depending on the promoter gene linked to the *lac* Z gene, different types of cells will therefore fluoresce when loaded with FDG. The distribution of these fluorescent cells (and therefore the activity of the specific promoter) can be visualized by looking at the intact embryo (Fig. 9.7). But the embryos can

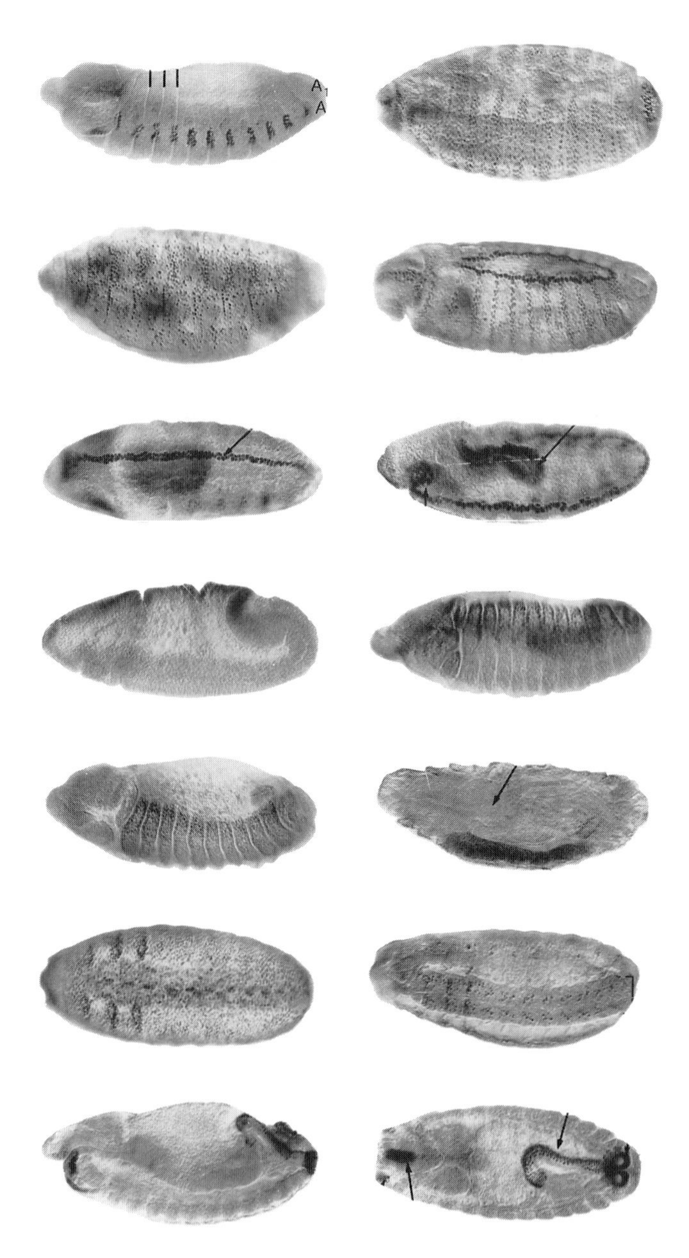

**Fig. 9.7.** *Drosophila* embryos transfected with the *lac* Z gene into association with different cell-type-specific promoter genes. Depending on the promoter gene, cells in different patterns over the embryo surface will possess the enzyme β-galactosidase (indicated by dark grains in this photograph). Courtesy of YN Jan from Bier et al. (1989).

also be dissociated and the cells expressing the reporter gene then sorted by flow cytometry, based on fluorescein fluorescence intensity (Fig. 9.8). In this way, cells destined for different functions can be purified and their subsequent development and interactions with other cells observed in culture (Fig. 9.9).

## MICROBIOLOGY AND GEL MICRODROPLETS

Although microorganisms would seem to be ideal candidates for flow analysis, flow cytometry has been slow to make its presence felt in the field of microbiology. To a great extent, this is attributable to limitations of the instrumentation; nozzles and sheath fluid and electronics designed for

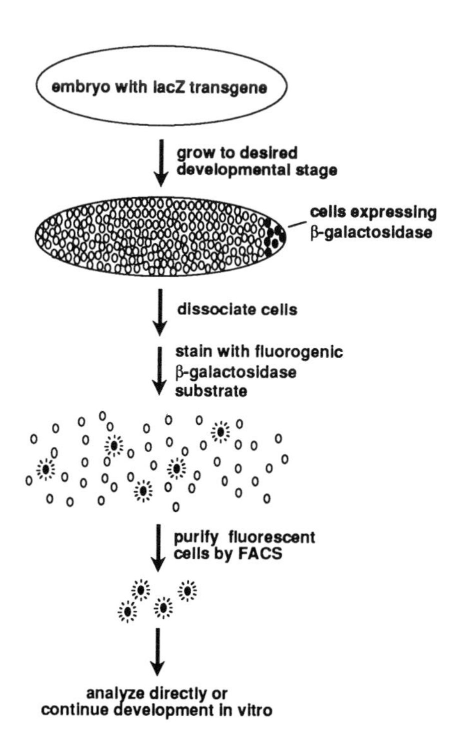


**Fig. 9.8.** The experimental protocol for whole animal cell sorting (WACS). From Krasnow et al. (1991).

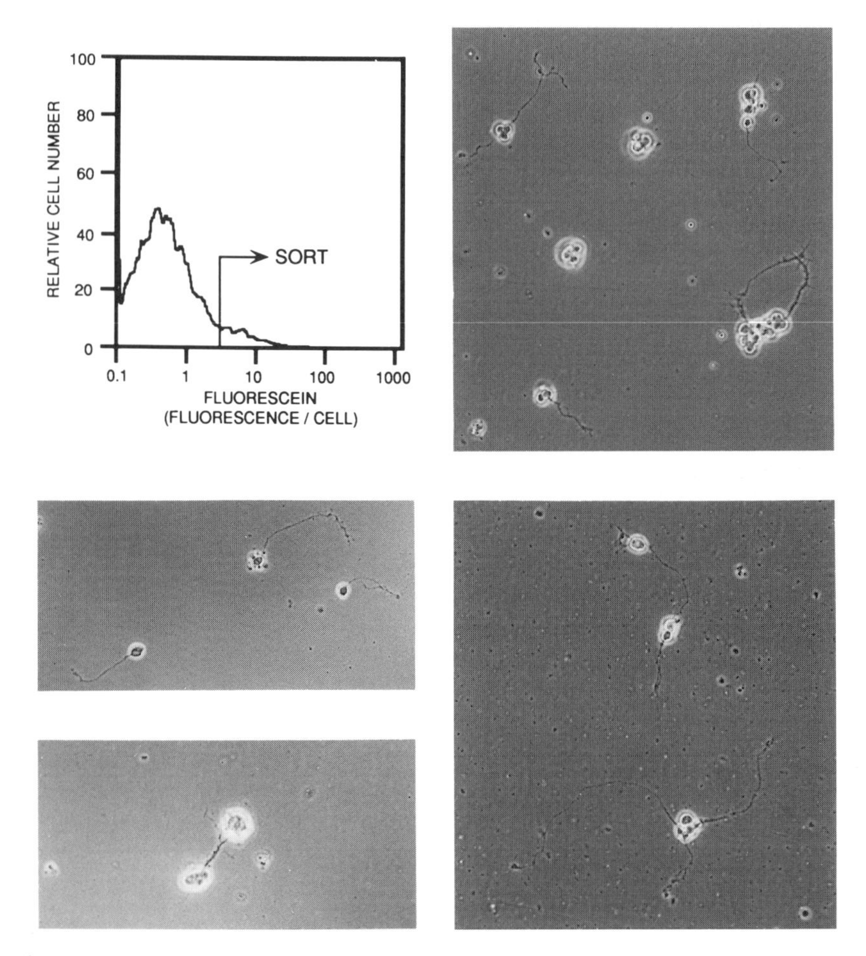


**Fig. 9.9.** When *lac* Z has been transfected into *Drosophila* embryos in association with a neuronal-cell-specific promoter, the cells that fluoresce brightly in the presence of fluorogenic β-galactosidase substrate will, when sorted, develop neuronal processes in culture. From Krasnow et al. (1991).

eukaryotic cells have not worked very well with the smaller members of the universe. The main problem has arisen from the difficulty in resolving low-intensity scatter and fluorescence signals from debris and instrument noise. A considerable amount of current work is aimed at overcoming some of these instrumental difficulties; and some cytometers perform better than others in this respect.

Much of the work on microorganisms in flow systems has concerned yeast, algae, and protozoa; although smaller than mammalian cells, these eukaryotes are considerably larger than most bacteria. DNA, RNA, protein, and light scatter measurements have been made on these organisms, and the feasibility of cell cycle analysis has been demonstrated. Bacteria, however, present more acute difficulties. The diameter of bacterial cells is perhaps 1 μm (compared with 10 μm for mammalian blood cells), and therefore the surface area to be stained (and resulting fluorescence intensity) is $10^2$ less than that of a mammalian cell. The DNA content of the *E. coli* genome is about $10^{-3}$ times that of a diploid human cell. Hence, bright dyes and sensitive instrumentation are required for studies of bacteria. Nevertheless, reasonable DNA histograms of bacteria can be obtained by flow cytometry. Methods are being developed to investigate cell cycle kinetics, the effects of antibiotics, and the detection and identification of bacteria for clinical investigations.

However, a technique developed at the Massachusetts Institute of Technology has approached the problem with the "if you can't beat them, join them" philosophy of allowing bacteria to masquerade as larger particles. The technique involves the creation of "salad oil" emulsions of drops of agar within bacterial suspensions in buffer solution. By adjusting the size of the drops and the concentration of the bacteria, it is possible to arrange conditions so that, on average, each drop of gel contains one bacterial cell. The microdrop then becomes a minicontainer for the bacterial cell, allowing the diffusion of stain, nutrient, antibiotics, and so forth, but containing the bacterial cell and all its progeny.

The usefulness of this technique is shown by its ability to detect the division of these bacterial cells. By staining the cells within the droplets in some way (*e.g.*, for DNA or protein content), the original culture will form a single flow histogram peak representing gel microdroplets, each fluorescing with an intensity related to the DNA or protein content of its entrapped single bacterial cell. After one replication cycle in which all the bacteria are replicating, each droplet will then contain two cells and have twice the original fluorescence intensity. Alternatively, if only some of the bacteria are replicating, a small population of gel droplets with twice the fluorescence intensity will appear. The droplets containing replicating cells will then progress to 4-fold, 8-fold, and 16-fold intensity as replication continues (Fig. 9.10). The technique can provide a sensitive method for studying small particles as well as a very rapid assay for the replication of a small proportion of bacterial cells in the presence of antibiotics, growth factors, or varied growth conditions (Fig. 9.11).

Although this gel microdroplet method is new and its potential applications relatively untested, it has been described here because it can teach us

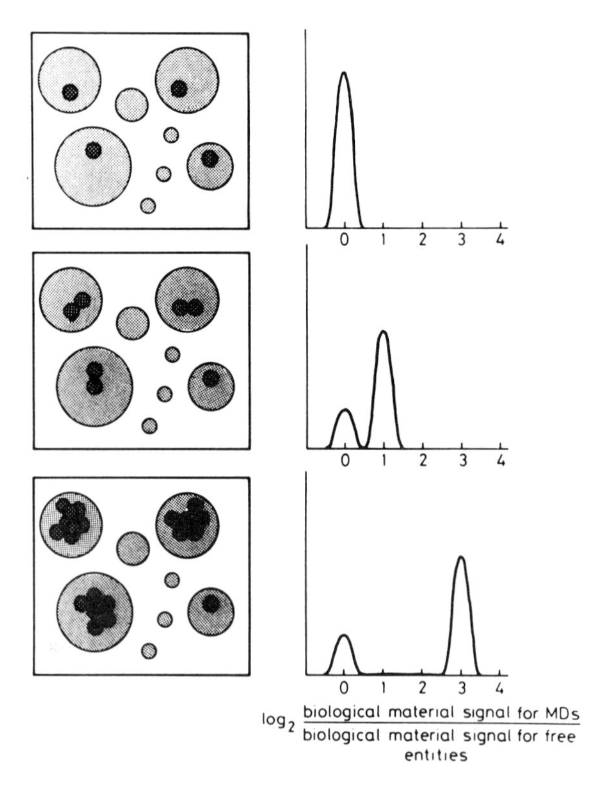


**Fig. 9.10.** Illustration of the use of gel microdroplets for sensing growth at the level of one cell growing into a two-cell microcolony. By staining the cells within microdroplets with, for example, a DNA-specific fluorochrome, a small subpopulation of cells dividing more or less rapidly than most could be detected in a flow histogram of microdroplet fluorescence. From Weaver (1990).

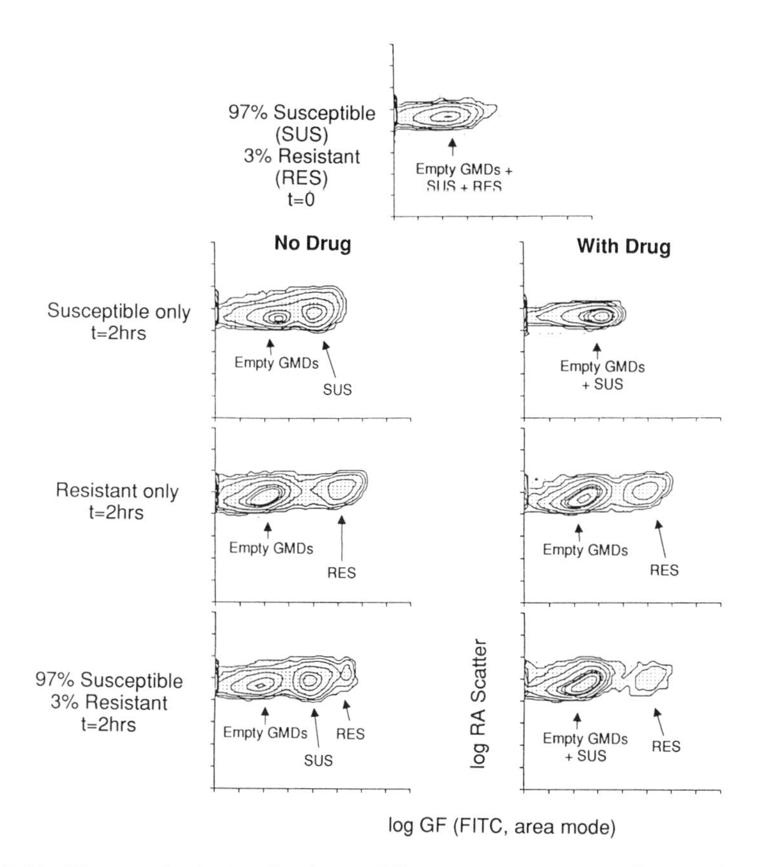


**Fig. 9.11.** The use of gel microdroplets and flow cytometry to assay drug sensitivity of bacterial cells. The figure shows side scatter and green fluorescence contour plots of gel microdroplets (GMDs) containing *E. coli* cells that have been stained with fluorescein isothiocyanate for total protein. The microdroplets have been analyzed in the flow cytometer either at time 0 or 2 hours after incubation in control medium (left plots) or medium containing penicillin (right plots). A model system was created by mixing two strains of bacteria (susceptible or resistant to penicillin). The data show that a small subpopulation of resistant cells could be detected within 2 hours because of its rapid growth in comparison to susceptible cells. From Weaver et al. (1991).

certain general lessons. It serves to remind us that cytometry, despite its name, does not necessarily involve the flow analysis of cells; particles of many sorts will do just as well. It is also of interest as a method that has, in fact, institutionalized the formation of clumped cells that most workers try so hard to avoid. In addition, it has provided us with a method for making a small cell into a larger (flow-friendly) particle. Finally, it has given us inspiration by exemplifying the way in which lateral thinking can extend the impact of flow cytometry in new directions.

## FURTHER READING

Chapter 32 in **Melamed et al.** is a good review of the use of calcium probes in flow cytometry. Grynkiewicz A, Poenie M, Tsien RY (1985). A new generation of $Ca^{2+}$ indicators with greatly improved fluorescence properties. J. Biol. Chem. 260:3440–3450 provides an entrance point into the research literature on the range of probes developed by Tsien's group.

A special issue of *Cytometry* (Vol 10, No. 5, 1989) is devoted to "Cytometry in Aquatic Sciences." In addition, Chapter 31 of **Melamed et al.** discusses the application of flow cytometry to higher plant systems.

Discussion of the WACS methodology can be found in Krasnow MA, Cumberledge S, Manning G, et al. (1991). Whole animal cell sorting of *Drosophila* embryos. Science 251:81–85.

The gel microdroplet technique is described in articles by Powell KT, Weaver JC (1990). Gel microdroplets and flow cytometry: Rapid determination of antibody secretion by individual cells within a cell population. In Bio/Technology 8:333–337; and Weaver JC, Bliss JG, Powell KT, et al. (1991). Rapid clonal growth measurements at the single cell level. Bio/Technology 9:873. Other work on flow cytometry of microorganisms is reviewed in Chapter 29 of **Melamed et al.**

# 10

# Flowing On: The Flow Cytometry Laboratory of the Future

In the Preface, I referred to the way flow cytometry has moved over the past two decades in directions that have surprised even those intimately involved in the field. It is obvious that attempts to predict the future of flow cytometry should be left to informal discussions between friends (preferably after a good dinner and a good bottle of wine); they should never be put down in print because they will probably turn out in time either to be wrong or else to be accomplished before the book has even been published. However, the temptation to speculate has been an impossible one to resist. This chapter should therefore be read in the spirit in which it was written: with a large measure of scepticism (preferably after a good dinner and a good bottle of wine).

The capabilities of state-of-the-art research cytometers should, over the next few years, move quite rapidly toward the goal of being able routinely to analyze particles of greater size range more sensitively and more rapidly than they do now. The size range will include particles from over 100 μm (like plant protoplasts) down to viruses, and perhaps DNA fragments and smaller molecules. This ability will come both from changes in design of the stream and nozzle and from increased electronic and optical sensitivity for the detection of dull signals. In practice, this will almost certainly lead to an increased use of flow cytometers for looking at plant cells and also at microbiological systems, including bacteria, viruses, and marine organisms. It may also lead to the routine use of cytometers to look at the properties of isolated organelles such as mitochondria and chloroplasts.

The speed with which these new instruments will carry out analysis will increase, both to keep up with demand by impatient scientists and, more importantly, to facilitate the sorting and analysis of rare particles. High-speed sorters at Los Alamos and Livermore National Laboratories currently make use of high-pressure flow (high stream velocity) to sort particles at the rate of about 30,000/sec. Higher rates than that (say, 100,000 particles per second) may be possible—given appropriate design of fluidics as well as optical and computer systems capable of keeping up with this rate. A second method for allowing more effective use of sorter time has been to sort four populations (instead of two) simultaneously. In the current design, this involves charging some drops more than others, so that drops can be deflected slightly to the left and to the right as well as strongly to the left and to the right—giving four options and allowing four populations to be sorted during each run. Four populations might not be the limit in the future.

The number of parameters measured per particle has been increasing over the years and is not limited so much by technical considerations as by intellectual ones. Whereas most instruments today look at FSC and SSC as well as three fluorescence parameters (*e.g.*, green, orange, and red), high-tech instruments already quite happily look at eight parameters (perhaps five fluorescences, FSC, SSC, and a Coulter volume signal), and, with demand, the electronics are capable of handling more. These extra parameters may include side scatter at different angles, polarization changes, and axial light extinction. In fact, there may well be more future interest in analysis of the intrinsic characteristics of cells that lead to variable patterns of light scatter and autofluorescence.

However, in a trend that is already quite apparent, extra parameters may also include ever-increasing numbers of simultaneous fluorescent stains. Systems with three or more lasers may become commonplace to allow the use of a wider range of fluorochromes; however, with advances in optics, there may also be a move back to arc lamps and to the spectral flexibility that they provide. The real developments will need to come from advancement in fluorochrome technology, which, in conjunction with extra lasers or arc lamps, will facilitate multiparameter analysis by allowing the use of wider regions of the spectrum. The question then will come to concern the ability of the human intellect to cope with the information that can be obtained in a short period of time from particles that are analyzed for eight or more parameters. It is not clear that our ability to design meaningful experiments that require large numbers of measured parameters has yet caught up with our ability to measure those parameters. One cannot help but notice with amusement that, despite the ability to look simultaneously at three fluorescence stains quite routinely on many instruments, most people seem quite happy to stick to two. Whether this is related to general conservatism, the high cost of additional antibodies, problems with cross-reactivity in complex staining systems, or some inherent human discomfort with forays into multidimensional reasoning is a question that may be answered in the future.

The extra information collected by rapid cytometers looking at eight or more parameters will certainly increase the demand on data storage systems. Flow cytometrists may begin to learn to wipe out data that they no longer need. A more likely alternative is that the need for new forms of data storage will push flow cytometry into the realm of optical discs for routine use. The trend may continue toward a proliferation of varied software packages for flow analysis; it is hoped that standard storage format will allow full compatibility of data acquired on any and all systems. We may, however, be reaching the limit on the pretty picture school of flow analysis: As flow plots become second nature to scientists, the macho value of star-studded, multicolored, multidimensional slides will decrease. However, some creative individual may still fill a need by coming up with a method to plot multiparameter data so that interpretation does not look like more trouble than it is worth.

An extra parameter that should gain more attention is that of time. Kinetics experiments will become more popular, as will the use of functional probes. There will also be an increase in the use of pulse processing (time-of-flight) analysis of signals to determine the shape characteristics of particles. Slit scanning technology may find its way into routine instruments in the future, thus facilitating the study of the detailed pattern of marker distribution within a cell or over the particle's surface. It may also happen that flow cytometrists and microscopists will begin, at last, to talk to each other when they discover that they share common interests in image analysis microscopy. A cytometry laboratory of the future may well be equipped with a collection of fluorescence microscopes, flow cytometers, and image analysis systems—all being used, as appropriate, by the same people.

I said at the beginning of this book that flow cytometry is currently moving in two directions at once: Technological advances provide, on the one hand, increasingly rapid, sensitive, complex, but precarious analysis and, on the other hand, increasingly simplified, fool-proof and automated capabilities. Ten years ago, all flow cytometry was at the complex, but precarious level of development. Now, many of those early precarious developments have been incorporated into routine, simplified cytometers, and new techniques are appearing in research instruments. Over the next few years, this progression will continue. But the balance has already begun to change so that the vast majority of future flow cytometric analysis will be done using routine, black-box instruments.

In association with this trend, expert (artificial intelligence) systems will begin to have an impact. Robotic systems, incorporating staining, lysing, and centrifugation steps, may allow staff to put blood or other crude material in at one end and get formatted print-outs of flow data from the other. Before this happens, there will need to be some advances in the use of flow cytometric controls. These may either be stained beads or fixed cells of some sort, but their routine use will be required if flow systems are going to be automated. Particularly in clinical and industrial laboratories, but in research laboratories as well, it is possible that the hands-off flow cytometer will be used increasingly in much the same way as today's Coulter counters and scintillation counters. It is likely that some type of flow cytometry will begin to be used in most microbiology and genetics laboratories. There can be little doubt that its use will become entrenched in the hematology, immunology, oceanography, transplantation, and pathology labs, where it has already had such impact. Flow cytometry may also begin to enter the biochemistry laboratory, where the use of monoclonal antibodies conjugated to beads makes it an alternative to the ELISA technique for assaying soluble components.

And what does the future hold for the *spirit* of flow cytometry? Although there will continue to be many new technical developments (some predicted and some surprising), I suspect that something of the pioneering mystique and sense of adventure will be lost from the field as the balance shifts from innovative technology to routine usage. But the trade-off for that loss of excitement will be the satisfaction we should feel from having played a part in the comfortable acceptance of this powerful technique by the broad scientific community.

# General References

Although articles on flow cytometry can be found throughout a great range of scientific publications, the following references are ones that I have found particularly useful for general information on the theoretical basis of flow analysis and as routes into the literature on particular subjects and techniques.

## BOOKS

Darzynkiewicz Z, Crissman HA (eds) (1990). Methods in Cell Biology, Vol 33, Flow Cytometry. San Diego: Academic Press. An up-to-date compilation of techniques covering most aspects of preparation and staining of cells for flow cytometry, but with an emphasis on nucleic acid and cell cycle analysis.

Grogan WM, Collins JM (1990). Guide to Flow Cytometry Methods. New York and Basel: Marcel Dekker, Inc. A straight-forward book with good coverage of practical protocols.

McMichael AJ, et al. (1990). Leukocyte Typing IV. Oxford: Oxford University Press. Tells more than I thought possible about antigens on the surface of leukocytes. It is reassuring to find out that even the experts have trouble keeping up with ever increasing CD numbers. New editions follow each of the frequent Leukocyte Typing Workshops.

Melamed MR, Lindmo T, Mendelsohn ML (eds) (1990). Flow Cytometry and Sorting, ed 2. New York: Wiley-Liss. A very thick (824 page) multi-author compendium, containing thorough review articles on the theory and practice of flow cytometry by acknowledged experts.

Ormerod MG (ed) (1990). Flow Cytometry, A Practical Approach. Oxford: IRL Press. A short book with some discussion of flow theory but with an emphasis on practical aspects of flow cytometry and technical protocols.

Robinson JP (ed) (1990). Handbook of Flow Cytometry Methods. West Lafayette, IN: Purdue University Cytometry Laboratories. An informal manual of techniques that are relevant to flow cytometric analysis. Inexpensive and well worth having (copies are available from ISAC, see below).

Shapiro HM (1988). Practical Flow Cytometry, ed 2. New York: Alan R. Liss, Inc. Manages to make learning about flow cytometry more enjoyable than you would have thought possible. There are readable sections on many aspects of flow theory and a good section on electronics for those who want to build their own....

Van Dilla MA, Dean PN, Laerum OD, Melamed MR (eds) (1985). Flow Cytometry: Instrumentation and Data Analysis. London: Academic Press. A venerable book with an emphasis on the physics and mathematics of flow systems and data analysis. It has some very readable articles on some rather theoretical subjects.

Watson JV (1991). Introduction to Flow Cytometry. Cambridge: Cambridge University Press. A somewhat idiosyncratic tour through the many theoretical aspects of flow cytometry with which Watson is well-acquainted. There are detailed discussions of the limits on signal resolution and on methods for looking at dynamic cell events. There is also good coverage of oncological applications but no mention at all of lymphocytes.

Weir DM (ed) (1986). Handbook of Experimental Immunology, Vol 1: Immunochemistry. Oxford: Blackwell Scientific Publications. A detailed reference volume (co-edited by the Herzenbergs) on many aspects of immunology, including immunofluorescence techniques and antibody conjugation methods as well as flow cytometric analysis.

## CATALOGUES

The following are catalogues and handbooks from manufacturers. They are either cheap or free and can be well worth reading.

Optics Guide 4 from Melles Griot (300 East River Road, Rochester, NY 14623; 1 Frederick Street, Aldershot, GU11 1LQ, UK). This guide is free and provides a great deal of information about filter, lens, and mirror specifications and design as well as about the Melles Griot range of high-quality products.

Molecular Probes: Handbook of Fluorescent Probes and Research Chemicals by Richard P. Haugland. Molecular Probes (PO Box 22010, 4849 Pitchford Ave, Eugene, OR 97402) makes a vast range of fluorescent chemicals that are useful in flow cytometric analysis. The Handbook provides a great deal of information about the use of these chemicals as well as about their photochemical characteristics.

Monograph: Fluorescent Microbeads Standards. This is a handbook printed by Flow Cytometry Standards Corporation (PO Box 12621, Research Triangle Park, NC 27709). It provides good information about the use of beads for calibration, compensation, and standardization of flow systems.

Hamamatsu Photomultiplier Tubes (360 Foothill Road, PO Box 6910, Bridgewater, NJ 08807). This catalogue contains a good discussion of photomultiplier tube electronics. We all need to be reminded occasionally that a flow cytometer's performance is never any better than the performance of its photodetectors.

Dako Handbook on Flow Cytometry (Dakopatts A/S Produktionsvej 42, DK-2600 Glostrup, Denmark). This heavy and attractive loose-leaf file from a company that sells monoclonal antibody reagents contains good information on staining techniques and on the CD markers for human blood cells.

## MISCELLANEOUS

The journal that specializes in research reports about flow techniques and flow analysis is *Cytometry*. The *Journal of Immunological Methods* is also often useful in this regard. *Cytometry* publishes special issues and supplements that focus on particular subjects like Aquatic Sciences and Cytogenetics. A subscription to *Cytometry* comes free with membership in the International Society for Analytical Cytometry (ISAC).

ISAC (PO Box 7849, Breckenridge, CO 80424) is the Society that specializes in flow and image cell analysis. Meetings (attended by about 1,000 people) are held approximately every 18 months (sometimes in the United States and sometimes in Europe).

The U.S. National Flow Cytometry Resource is at the Los Alamos National Laboratories (Los Alamos, NM 87545). They are a source of information and provide help and facilities for scientists wanting to make use of their "state of the art" cytometers. They also run courses and publish a free newsletter, which can be obtained by contacting Mary Cassidy, Editor (Mail Stop M888, at the above address).

# Glossary

There is a fine line between words that provide necessary technical information and words that we might call *jargon*. Whereas technical vocabulary is important as a means to intellectual precision, jargon can often be used either to obscure ignorance or, like a badge, to identify members of an exclusive club. Both technical vocabulary and jargon, however, form a barrier between people already within a field of endeavor and those attempting to enter that field. Without identifying which of the following words are necessary and which are merely jargon, I include this somewhat selective glossary as an effort toward lowering that barrier.

**Absorption:** In the context of photochemistry, *absorption* refers to the utilization, by an atom or molecule, of light energy to raise electrons from their ground state orbitals to orbitals at higher energy levels. Having absorbed the light energy, the atom or molecule is now in an excited state and will emit energy (in the form of either heat or light) when it returns to its ground state. Atoms will absorb light if, and only if, it is of a wavelength whose photons contain exactly the amount of energy separating a pair of electron orbitals within that atom.

**Acquisition:** In flow cytometry, *acquisition* refers to the process of recording the intensity of the photodetector signals from a particle in the transient memory of a computer. Once acquired, the data from a group of particles can be stored permanently on a storage medium from which it can be subjected to analysis. Acquisition and then analysis (in that order) are the two central steps in the flow cytometric procedure.

**Acridine orange:** Acridine orange (AO) is a stain that fluoresces either red or green, depending on whether it is bound to double-stranded or single-stranded nucleic acid. It has proven useful in comparing DNA and RNA content within cells; and it has also been used successfully (in the presence of RNAase and mild denaturing conditions) to look at the changes in DNA denaturability during the cell cycle. There is debate in the flow cytometric community about whether the reputation that AO has for being difficult to work with is justified.

**ADC:** An analog to digital converter (ADC) assigns light intensity ranges to each of its 256 or 1024 channels and then converts the electrical signal from a photodetector (called an *analog signal* because it is continuously variable and can have an infinite variety of values) into a value that is merely the number of a specific channel, depending on the intensity range of which channel the signal's value falls within.

**Aerosol:** An aerosol is the spray of small fluid droplets that can be generated particularly when the nozzle of a flow cytometer is vibrated for sorting applications. If samples contain material that may be a biological hazard, attention should be paid to containment of the aerosol by suction through small pore, hydrophobic filters.

**Algae:** Algae are simple forms of plant life. The larger algae are known as *seaweeds*. The unicellular algae form a large part of the plankton of both marine and freshwater environments and are suitable for analysis by flow cytometers. Algae also have contributed greatly to flow cytometric analysis because of their elaboration of pigments like phycoerythrin and allophycocyanin, which can be conjugated to antibodies and which facilitate multicolor staining procedures because they fluoresce in different regions of the spectrum.

**Analysis:** After acquisition, data from a sample are processed so as to provide useful results. This processing, or analysis, consists merely of the correlation, in some or all possible ways, of the intensity channel from each of the signals recorded for each of the particles acquired for a given sample.

**Analysis point:** The three-dimensional point in space where the laser beam intersects and illuminates the sample core of the fluid stream is the analysis point. The size of this point is determined by the cross-sectional dimensions of the laser beam and the width of the stream core itself. It is within the volume of the analysis point that particles are illuminated.

**Aneuploid:** Although *aneuploid* is used by cytogeneticists to refer to cells with abnormal numbers of chromosomes, it has been hijacked (with intellectual imprecision) by flow cytometrists to refer to the characteristic of possessing an abnormal amount of DNA.

**Arc lamp:** An arc lamp is a device consisting of a filament within a glass bulb that emits light of a wavelength determined by the material of the filament.

**Autofluorescence:** The light emitted naturally by an unstained particle is called autofluorescence. The amount of autofluorescence will differ depending on the type of particle, on the wavelength of the illuminating beam, and on the fluorescence wavelength being analyzed. Autofluorescence, in general, results from endogenous compounds that exist within cells. It can be studied as an interesting phenomenon in itself; however, bright levels of autofluorescence at a particular wavelength will lower the sensitivity of a flow system for detecting positive stain of low intensity.

**Axial light loss:** Axial light loss is a parameter that can be measured in some flow systems and is related broadly to the physical characteristics of a particle. It is defined as the total amount of light that is lost from the laser beam by virtue of its striking a particle within the flow stream. The light loss results from a combination of factors including light absorption, light refraction, and light scatter.

**Back-gating:** Back-gating is a strategy by which a mixed population of cells that have been stained with an antibody specific for a subset of that population are analyzed to determine the side scatter and forward scatter characteristics of the stained cells. This is in contrast to "traditional" gating, whereby the known scatter characteristics of a class of cells are used to delineate cells whose staining characteristics are then determined.

**Beads:** Beads are particles (made, usually, of latex or some form of polystyrene) that can be used as stable and inert standards for flow cytometric analysis. Beads can be obtained in different, narrowly defined, sizes in order to standardize the FSC settings. They can also be obtained conjugated to various fluorochromes in order to standardize the fluorescence detection settings.

**Bit map:** A two-dimensional diagram used to display particles in order to correlate the intensities for each particle of two flow cytometric parameters. *Dot plot* is a synonymous term. Bit maps suffer, graphically, from black-out (or green-out [or white-out]) if you are looking at a computer screen) in that an area of a display can get no darker than completely black (or green [or white]); if the number of particles at a given point is very dense, their visual impact, in comparison with less dense areas, will decrease as greater numbers of particles are displayed. Contour plots display the same kind of correlation as bit maps and dot plots but, because the levels of the lines can be altered, can provide more visual information about the density of particles at any given point in the correlation display.

**BrdU:** 5-Bromodeoxyuridine (also abbreviated BUdR or BrdUrd) is a thymidine analog that will be incorporated into the DNA of cycling cells. Cells pulsed with BrdU can then be stained with anti-BrdU monoclonal antibodies to indicate which cells have been synthesizing DNA during the pulse period.

**Break-off point:** The point, some distance from a nozzle, at which a vibrating stream begins to separate into individual drops that are detached and electrically isolated both from each other and from the main column of the stream.

**Calcium:** An ion whose rapid flux between cellular compartments and from the external medium is thought to be important in signal transduction within cells. Its concentration can be measured by various fluorescent probes, such as fluo-3 and indo-1.

**Channel:** Channel is the term by which a flow cytometer characterizes the intensity of the signals emitted by a particle. Most cytometers divide the intensity of light signals into either 256 or 1024 channels. Signals with high channel numbers are brighter than signals with low channel numbers; however, the quantitative relationship between signals defined by one channel number and those defined by another will depend on the amplifier and photodetector voltage characteristics of a given protocol.

**Coaxial flow:** The flow of a narrow core of liquid within the center of a wider stream. Flow of this type is important in flow cytometry because it provides a means by which particles flowing through a relatively wide nozzle can be tightly confined in space, allowing accurate and stable illumination as they pass one by one through a light beam.

**Coherence:** Coherence is the property of light emitted from a laser such that it is remarkably uniform in color, polarization, and spatial direction. Spatial coherence allows a laser beam to maintain brightness over a great distance.

**Compensation:** Compensation is the ability of a flow cytometer to correct for the overlap between the fluorescence spectra of different fluorochromes. Without compensation, fluorescence from a given fluorochrome may register to some extent on a photodetector assigned to the detection of a different fluorochrome.

**Contour plot:** A contour plot is one method for displaying data correlating two cytometric parameters so that the density of particles at any place on the plot is used to generate contour lines (much as contour lines on a topographic map are used to describe the height of mountains at different points on the landscape). The particular algorithm used to generate the contour lines can make a great difference in the visual impact of the display. See also **bit map.**

**Control:** A control is a type of sample of which, most particularly in flow cytometry, you generally have one fewer than you actually need.

**Core:** The core is the stream-within-a-stream that has been injected into the center of the sheath stream and is maintained there by the hydrodynamic considerations of laminar flow. The core contains the sample particles that are to be analyzed in the flow cytometer.

**Coulter volume:** The increase in electrical resistance that occurs as a particle displaces electrolyte when it flows through a narrow nozzle is the particle's Coulter volume. This increase in resistance is only roughly related to the volume of the particle.

**Cross-match:** Cross-matching is the process of testing the cells of a prospective organ donor with the serum of a prospective organ recipient for compatibility. The flow cytometric cross-match determines whether serum of the recipient contains antibodies that bind to the donor cells. Such binding constitutes a positive cross-match and is a contraindication to transplantation in that particular donor/recipient combination.

**Cross-talk:** Cross-talk is the signal from the "wrong" photodetector that results because the fluorescent light emitted by one fluorochrome contains some light of a wavelength that gets through the filters on the photodetector that is nominally specific for the fluorescence from a different fluorochrome. See **compensation**.

**CV:** The coefficient of variation (CV) is defined as the standard deviation of a series of values divided by the mean of those values. It is used in flow cytometry to describe the width of a histogram peak. Whereas in some protocols it can be used to assess the variation in particle characteristics within a population, in DNA analysis (where all normal particles are assumed to have identical characteristics) it is frequently used to assess the alignment of a flow cytometer (and the skill of its operator). Discussions about CV have been known to bring out rather primitive competitive instincts within groups of flow cytometrists, who are usually friendly, well-adjusted people.

**Cytofluorimeter:** *Cytofluorimeter* is a term synonymous with flow cytometer, but with slightly antiquated overtones.

**Deflection plates:** Deflection plates are the two metal plates, one on each side of the flow stream, that carry a high voltage used to deflect drops that have been charged either negatively or positively to one side or the other of the main stream. It is by means of the charge on the deflection plates that sorting of particles occurs. The deflection plates also provide a good reason for flow cytometry operators to keep their hands dry and wear rubber-soled shoes.

**Dichroic mirror:** A mirror that by virtue of its coating reflects light of certain wavelengths and transmits light of other wavelengths is a dichroic mirror. Such mirrors are used in flow cytometry to direct light from specific fluorochromes toward particular photodetectors.

**Dot plot:** See **bit map**.

**Drop delay:** Drop delay is the time (often measured in number of drops, because the number of drops per second is known) between the measurement of the signals from a particle and the moment when that particle is just about to be trapped in a drop that has broken off from the main column of a vibrating stream. It is at this precise moment that the column of the stream must be charged (and then shortly thereafter grounded) if the drop containing a particle of interest is to be correctly charged and then deflected to the left or right of the main stream.

**Dual-parameter correlated mode:** See **list mode data**.

**Emission:** Emission is the loss of energy from an excited atom or molecule in the form of light. Although a long-lived form of light emission (known as *phosphorescence*) does occur, in flow cytometry the light emission that we are concerned with occurs rapidly after excitation and is called *fluorescence*.

**Erythrocyte:** The red, hemoglobin-containing cells that occur in the peripheral circulation and are responsible for transporting oxygen are erythrocytes. Despite this vital physiological function, flow cytometrists are apt to feel a certain amount of hostility toward erythrocytes because they outnumber white cells by about 1000 to 1 and therefore make analysis of white cells difficult unless removed by either lysis or density gradient centrifugation. Fixed erythrocytes from chickens are an exception to this general hostility. When fixed in glutaraldehyde, chicken erythrocytes fluoresce brightly at a broad range of wavelengths and are therefore a useful tool for aligning the cytometer beam; when fixed in ethanol, they are a useful DNA standard (containing approximately one-third the amount of DNA of a normal human cell).

**Euploid:** Flow cytometrists use the term *euploid* to refer to a cell with the "correct" amount of DNA for its species. Because its sensitivity is limited and because it works only with total DNA content, a flow cytometer may not agree with a microscopist (who looks at chromosomes) in the classification of cells by this criterion. See also **aneuploid**.

**FACS:** FACS is an acronym for fluorescence-activated cell sorter. It is a term coined by Becton Dickinson for its instruments, but has come to be used generally as a term that refers to all instruments that analyze the light signals from particles flowing in a stream past a light beam. The term *flow cytometer* is perhaps more correct since it has neither the trademark connotations nor any reference to a sorting function, which most of these instruments no longer possess. The term *cytofluorimeter* is also used, but generally has antiquated overtones.

**FCS format:** FCS is a standard format for flow cytometric data storage to facilitate the programming for analysis of results. Manufacturers have been regrettably slow to embrace this standard fully.

**Filter:** A filter is a piece of glass that modifies the light that passes through it. A neutral density filter reduces the intensity of a light beam without affecting its color. Short-pass, long-pass, and band-pass filters selectively alter the color of a light beam by transmitting light of restricted wavelength.

**Fixation:** A process by which the protein of cells is denatured. Fixation in flow cytometry is used to inactivate hazardous biological material and also to preserve stained cells when there is not immediate access to a flow cytometer. Paraformaldehyde is the fixative of choice for flow cytometry because it preserves the forward and side scatter characteristics of cells (but causes some increase in their autofluorescence).

**Flourescence:** A misspelling of *fluorescence* often found in keyword listings for articles on flow cytometry.

**Flow:** *Flow* is a term that has come to refer, colloquially, to the general technique of flow cytometry. For example "Do you use flow to analyze your cells?" or "Flow has revolutionised the study of immunology."

**Flow cell:** The flow cell is the device in a flow cytometer that delivers the sample stream to the center of the sheath stream. In a sorting instrument, the flow cell vibrates in order to allow drop formation, and it also provides the electrical connections for charging and then grounding the stream at appropriate times. In some cytometric configurations, the laser illuminates the stream within the flow cell; in other configurations the illumination occurs "in air" after the stream has left the flow cell. In the latter case, the term *nozzle* is more apt to be used.

**Flow cytometry:** See Chapters 1 through 10 of this book.

**Flower:** *Flower* is a term of affection in the Northeast of England—and one that could, with a slight change of pronunciation, be adapted to refer to a person who works in the field of flow cytometry.

**Fluorescein:** Fluorescein is a fluorescent dye that can be readily linked to proteins and that is therefore useful, when conjugated to specific antibodies, for lighting up cells with particular phenotypes.

**Fluorescence:** Fluorescence is a form of light emitted by atoms or molecules when electrons fall from excited energy levels to their lower, less-energetic ground state. *Fluorescence* is often misspelled as *flourescence*, so it pays to check both spellings when doing a keyword literature search.

**Fluorochrome:** A fluorochrome is a dye that absorbs light and then emits light of a different color (always of a longer wavelength). Fluorescein, propidium iodide, and phycoerythrin, for example, are three fluorochromes in common use in flow cytometry.

**Fluorogenic substrate:** A fluorogenic substrate is a chemical that is an enzyme substrate and that becomes fluorescent when processed by that enzyme. A fluorogenic substrate can therefore, by virtue of its increasing fluorescence, be used to measure the activity of a given enzyme within a cell.

**Forward scatter:** Forward scatter is light from the illuminating beam that has been bent (refracted or otherwise deflected) as it passes through a particle so as to diverge from the original direction of that beam. The intensity of the light bent to a small angle from the illuminating beam is related to the refractive index of the particle as well as to its cross-sectional area. The forward scatter signal causes a great deal of confusion because some people who call it a *volume* signal actually begin to think that it is closely correlated with a cell's volume. *Forward scatter* is often abreviated as FSC or as FALS (for "forward angle light scatter").

**G0/G1:** In flow cytometry, the term refers to cells that have the 2C (diploid) amount of DNA and are therefore not cycling (G0) or are just completing cytokinesis and have not yet begun again to make more DNA in preparation for a new cycle (G1).

**G2/M:** In flow cytometry, the term refers to cells that have the 4C (tetraploid) amount of DNA and are therefore either just completing DNA synthesis in preparation for mitosis (G2) or are involved in the steps of mitosis prior to cytokinesis (M).

**Gain:** Gain is the electronic control on an amplifier that determines the current intensity that results when a given signal is received by a photomultiplier tube. Variation in the gain on photomultiplier tube amplifiers will vary the appearance of the output signals as they are converted into flow cytometric data.

**Gate:** Gate is a restriction placed on the flow cytometric data that will be included in subsequent analysis. A *live gate* restricts the data that will be accepted by a computer for storage; an *analysis gate* simply excludes certain stored data from a particular analysis procedure. A gate is used to restrict analysis of a mixed population to certain cells within that mixed population. It is most frequently used to ensure that cells with the FSC and SSC characteristics of lymphocytes are the only cells that are then analyzed for their staining characteristics. See also **back-gating**.

**Granularity:** Granularity is a term used synonymously with *side scatter* to describe the light that is deflected to a right angle from the illuminating beam in a flow cytometer. The intensity of this light is related, in an imprecise way, to internal or surface irregularities of the particles flowing through the beam.

**Granulocyte:** Granulocytes are a class of white blood cells (also called *polymorphonuclear cells*) that possess irregular nuclei and therefore have bright side scatter signals.

**Hard disk:** A hard disk is a storage medium for data. Hard disks have relatively large memory capacity but, when not removable, require further back-up facilities because they will always become filled faster than you think they will.

**Heath Robinson:** *Heath Robinson* is the English term for a complex instrument that looks as if it has been designed by a committee. A Heath Robinson device usually works well but is often held together by chewing gum and string. In the context of flow cytometry, this term requires no further explanation. See also **Rube Goldberg**.

**High-speed sorting:** High-speed sorting is a flow cytometric technique whereby adaptations to pressure and fluid controls that allow high stream velocities are made in order to increase the speed of sorting cells. Such adaptations are important when large numbers of relatively rare particles are required and have been particularly useful in the compilation of DNA libraries from sorted chromosomes.

**Histogram:** A histogram displays data from one parameter of flow cytometric data at a time.

**Hydrodynamic focusing:** Hydrodynamic focusing is the property of laminar flow that maintains particles that are in the central core of a column of fluid within that central core.

**Interrogation point:** *Interrogation point* is synonymous with **analysis point**.

**Laser:** The term is an acronym for light amplification by stimulated emission radiation. Lasers are important in flow cytometry because, as a result of their coherent output, they are a means of illuminating cells with a compact, intense light beam that will produce fluorescence signals that are as bright as possible over a short time period.

**Lens:** A lens is a means of changing the shape of a beam of light. In flow cytometry, lenses are used to narrow the laser beam to a small point at the stream. Some lenses produce a beam with a circular cross-sectional shape; others produce beams with an elliptical configuration. Lenses are also used in a flow cytometer to collect diverging light and to transmit it to an appropriate photodetector.

**Linear amplifier:** A linear amplifier is one means of increasing the signal from a photomultiplier tube to make it measurable. A linear amplifier increases the signal in such a way that the output current from the amplifier is directly proportional to the input current derived from the photodetector. See also **logarithmic amplifier**.

**List mode:** List mode is a method of data storage such that the intensities of the signals from each photodetector that are generated by a particle are stored in association with each other so that they act as a total flow cytometric description of that particle and can, for analysis, be correlated with each other (and with the descriptive signals from all other particles) in all possible ways. Dual-parameter correlated data storage and single-parameter data storage are, by contrast, methods of data storage that collect all the information from each photodetector for the total number of particles in a sample and sum it before storage, thus losing the total description of each particle and therefore limiting the potential for correlation between parameters during analysis. List mode data storage is always the option of choice, unless considerations of data storage capacity predominate.

**Logarithmic amplifier:** Logarithmic amplification is one means of modifying the signal from a photomultiplier tube to make it measurable. A logarithmic amplifier modifies the signal in such a way that the output current from the amplifier is in proportion to the logarithm of the input current derived from the photodetector. See also **linear amplifier**.

**Lymphocyte:** A lymphocyte is a particular type of white blood cell that is involved in many of an organism's immune responses. Subpopulations of lymphocytes with microscopically identical anatomy can be distinguished because their surface membranes contain different arrays of proteins. The staining of these proteins with fluorescently tagged monoclonal antibodies allows the subpopulations to be enumerated by flow cytometry.

**Marker:** Marker is a term that is often used to refer to a dividing line applied to a fluorescence intensity histogram in order to dichotomize particles into those that are to be called positively stained from those to be called unstained. Somewhat confusingly, it is also used by immunologists to refer to a significant protein on the surface of a cell.

**Mean:** Mean is a value that can be used to describe the fluorescence intensity of a population of cells. If the mean is calculated by summing the channel numbers from all the cells in a sample and then dividing by the total number of cells in that sample, then it will underrepresent the contribution made by the brightly fluorescent cells if a logarithmic amplifier has been used (see also **median** and **mode**). In addition, the calculation of a mean from a flow histogram will incorrectly evaluate any particles that lie in the highest and lowest channels.

**Median:** Median is a value that can be used to describe the fluorescence intensity of a population of cells. If the cells were lined up in order of increasing intensity, the median value would simply be the channel number of the cell that is at the midpoint in the sequence. The median channel is, for several reasons, the best way to describe the intensity of a population (see also **mean** and **mode**).

**Mode:** Mode is a value that can be used to describe the fluorescence intensity of a population of cells. The mode intensity is the channel number that has been used to describe the largest number of cells in a sample. Mode values are apt to be variable when intensity distributions are broad or when few particles have been analyzed (see also **mean** and **median**).

**Monocyte:** Monocytes are a class of white blood cells that copurify with lymphocytes in commonly used density gradient procedures, that tend to be promiscuously sticky for the nonspecific ends of monoclonal antibodies, and that therefore can lead to misleading results in analysis of lymphocyte subpopulations. Monocytes differ from lymphocytes in their FSC and SSC characteristics.

**Nozzle:** See **flow cell**.

**Obscuration bar:** An obscuration bar is a strip of metal or other material that serves to block out direct light from the illuminating beam. Any light reaching a photodetector will therefore be light that has been deflected around the bar by the physical characteristics of a particle interacting with the light. The width of the bar will define the narrowest angle by which the light must be deflected in order to reach the detector.

**Observation point:** See **analysis point**.

**Optical disk:** An optical disk is a high-capacity storage medium that may for a short while (until our addiction to data catches up) solve some of the problems generated by the need for large volumes to store flow cytometric information.

**Optical bench:** The stable table (!) that keeps the light beam, fluid streams, lenses, and photodetectors of a flow cytometer all precisely aligned with each other is the optical bench. Lack of stability in these components leads to artifactual results.

**Orifice:** The orifice is the exit hole in a flow cell or nozzle. Cells anywhere near as large as this hole will block it.

**Parameter:** Parameter is the term applied to the light signals measured by a flow cytometer. The number of parameters measured by a cytometer is determined by the number of photodetectors present; each photodetector, by virtue of the filters in front of it, is responsible for the conversion of a particular color of light into an electrical signal. Modest cytometers measure three parameters. Immodest cytometers measure eight parameters. Average cytometers measure between four and six parameters. In the future, our standards may increase.

**Peak reflect:** The peak reflect method is an algorithm for analyzing DNA histograms to determine the number of cells in the S phase of the cell cycle. In the peak reflect method, the shapes of the 2C and 4C peaks are assumed to be symmetrical, thus allowing subtraction of the contribution from these two peaks from the S phase cells between them.

**Photodetector:** A photodetector is a device that senses light and converts the energy from that light into an electrical signal. Within the operating range of the detector, the intensity of the electrical signal is proportional to the intensity of the light. Photomultiplier tubes and photodiodes are two types of photodetectors.

**Photodiode:** A photodiode is a type of photodetector used to detect relatively intense light signals. It does not have a high voltage applied to increase the current flow at its anode (output) end.

**Photomultiplier tube:** A photomultiplier tube is a type of photodetector used to detect relatively weak light signals. Its output current is increased by means of high voltage applied.

**Phycoerythrin:** Phycoerythrin is a fluorochrome derived from deep sea algae. It is particularly useful in flow cytometric applications requiring dual-color analysis because, like fluorescein, it absorbs 488 nm light from an argon laser. However, it has a longer Stokes shift than fluorescein, and therefore the fluorescences of the two fluorochromes can be distinguished.

**Probe:** Probe is a general term used, in flow cytometry, to refer to any chemical that fluoresces when it reacts or complexes with a specific class of molecules and therefore can be used to assay that molecule quantitatively. Propidium iodide and acridine orange are nucleic acid probes because they complex specifically with nucleic acids and fluoresce brightly when they have reacted in this way. Fluo-3 is a calcium probe because it chelates calcium ions and fluoresces brightly when it is complexed with this ion.

**Propidium iodide:** Propidium iodide is a probe that can be used to measure quantitatively the amount of double-stranded nucleic acid that is present in a cell. In the presence of RNAase, it will measure the amount of DNA present. Because it does not cross an intact cell membrane, cells need to be treated with detergent or ethanol before it can be used to determine their DNA content. It can also be used to assess the viability of cells.

**Quadrant:** For dual color analysis, a two-dimensional contour plot or bit map of particles according to the fluorescence intensities of two colors is, traditionally, divided into four quadrants (the division is based on the background fluorescence of the unstained control sample). The quadrants, from this division, will then contain 1) cells that have stained with the first fluorochrome only; 2) cells that have stained with both fluorochromes; 3) cells that remain unstained; and 4) cells that have stained with the second fluorochrome only. In other words, the quadrants are defined so that they delineate the two types of single positive cells (1 and 4), double-negative cells (3), and double-positive cells (2).

**Region:** A region is any group of particles that have been selected by means of their flow cytometric parameters. A region, in the old days, was always rectangular—with an upper and lower channel limit for each parameter described. Now, a region can take any shape, based on the channel numbers defining a given number of vertices of an irregular figure.

**Rube Goldberg:** Rube Goldberg is the American synonym for Heath Robinson.

**S-FIT:** An S-FIT approximation is a mathematical algorithm for guessing at which cells in a DNA-content histogram are actually in the S phase of the cell cycle. The S-FIT algorithm bases this guess on the shape of the DNA histogram in the middle region between the G0/G1 and the G2/M peaks.

**S phase:** S phase is that phase of the cell cycle during which cells are in the process of synthesizing DNA in preparation for cell division. During this phase, cells have between the 2C amount of DNA normal to their species and the 4C amount of DNA, which is exactly double the 2C amount. It is the overlap of fluorescence intensity between cells in S phase and some of the cells with the 2C and 4C amounts of DNA that leads to uncertainty in the flow cytometric estimation of S phase.

**Sheath:** Sheath is the fluid within which the central sample core is contained during coaxial flow from or within the flow cell of a flow cytometer.

**Side scatter:** Side scatter is light of the same color as the illuminating beam that bounces off particles in that beam and is deflected to the side. The "side" is usually defined by a lens at right angle (orthogonal orientation) to the line of the laser beam. Side scatter light (SSC) may alternatively be called *right angle light scatter* (RALS) or *90° LS*. The intensity of this light scattered to the side is related in a general way to the roughness or irregularity or granularity of the surface or internal constituents of a particle.

**Signature:** Signature is a term used by marine flow cytometrists to refer to the flow cytometric characteristics of a sample of water. The signature of a water sample is related to the forms of plankton found within that sample.

**Single-parameter mode:** See **list mode**.

**Slit scanning:** Slit scanning is a technique by which the laser beam or the signal from that beam is directed through a narrow slit (<1 μm) so that many individual time-separated signals are obtained from each single (relatively large) particle as it flows past the source of illumination. In this way, information can be obtained about variation of structures along the length of that particle. Slit scanning shows particular promise in the analysis of chromosomes for centromeric position but may also be useful for looking at nuclear shape within the cytoplasm of an intact cell.

**SOBR:** SOBR is a mathematical algorithm for guessing at which cells in a DNA-content histogram are actually in the S phase of the cell cycle. The SOBR (sum of broadened rectangles) algorithm bases this guess on the use of gaussian-broadened rectangular distributions to attempt to fit the shape of the DNA fluorescence intensity histogram of the cells in question.

**Stokes shift:** The Stokes shift is the difference (in either energy or wavelength) between the light absorbed by a fluorochrome and the light emitted when that fluorochrome fluoresces.

**Tape back-up:** A tape back-up is a computer operation for making a copy of flow cytometric data onto relatively cheap data cartridge tapes. Back-up tapes help to postpone the day on which a flow cytometrist will need to make difficult decisions about wiping out old data. See also **optical disks**.

**Tetraploid:** Tetraploid is a term used to describe cells with double the amount of DNA normal for a particular species. These cells may either be cycling cells in the G2 or M phases of the cycle or, alternatively, may be abnormal cells with the "wrong" amount of DNA for their species. Malignant cells are frequently of the tetraploid type. Distinguishing this type of aneuploidy from normal G2 or M cells or, indeed, from clumps of two cells can be difficult with flow cytometry.

**Threshold:** The threshold is an electronic device by which an ADC can be made to ignore signals below a certain intensity. A forward scatter threshold is most commonly used in flow cytometry to exclude very small particles, debris, and electronic or optical noise from acquisition.

**Time:** Time is a parameter that is being used more frequently in conjunction with flow cytometric analysis. It is being used both with the slit scanning technique and in relation to the functional analysis of cells for the rate of reaction with various probes and fluoregenic substrates.

**Volume:** Volume is a useful and definite characteristic of any particle—but one that is not amenable to flow cytometric analysis. Coulter volume and forward scatter do not measure volume.

**Wavelength:** A wavelength is a characteristic of light that is related exactly to its energy content and also (with light to which our eyes are sensitive) to its color. Light of short wavelength has more energy than light of longer wavelength (and lies toward the blue region of the spectrum). Wavelength is used to describe the characteristics of filters and of dichroic mirrors. In sorting flow cytometry, it is a term also used to describe the distance between drops as they form from a vibrating fluid jet. This drop wavelength is determined by the diameter of the stream.

# Figure Credits

Fig. 1.1. Reproduced with permission from Alberts B, et al. (1989). Molecular Biology of the Cell, 2nd edition. New York: Garland Publishing.

Fig. 1.2. Reproduced with permission from Kamentsky LA, Melamed MR (1967). Spectrophotometric cell sorter. Science 156:1364-1365. © 1967 by the AAAS.

Fig. 1.3. Photograph from the Lawrence Livermore National Laboratory, operated by the University of California under contract to the United States Department of Energy.

Fig. 1.4. Photograph by Edward Souza for the Stanford University News Service.

Fig. 1.5. Photographs from Becton Dickinson Immunocytometry Systems and Coulter Electronics.

Fig. 3.1. Adapted from Becton Dickinson Immunocytometry Systems, San Jose, CA.

Fig. 3.3. Reproduced with permission from Carter NP, Meyer EW (1990). Introduction to the principles of flow cytometry. Ormerod MG (ed). Flow Cytometry: A Practical Approach. Oxford: IRL, pp 1-28.

Fig. 3.4. Reproduced with permission from Pinkel D, Stovel R (1985). Flow chambers and sample handling. Van Dilla MA, et al. (eds). Flow Cytometry: Instrumentation and Data Analysis. London: Academic Press, pp 77-128.

Fig. 3.7. Reproduced from Blakeslee A (1914). Corn and men. J Hered 5:512.

Fig. 4.6. Reproduced from Dean, PN (1990). Data processing. In Melamed MR, et al. (eds). Flow Cytometry and Sorting. New York: Wiley-Liss, pp 415-444.

Fig. 5.3. Reproduced with permission of Spectra Physics.

Fig. 5.6. Reproduced with permission from Shapiro H (1988): Practical Flow Cytometry. New York: Alan R. Liss.

Fig. 6.1. Reproduced with permission from Kessel RG, Kardon RH (1979). Tissues and Organs: A Text-Atlas of Scanning Electron Microscopy. San Francisco: WH Freeman & Co.

Fig. 6.4. Reproduced with permission from Horan PK, et al. (1986). Improved flow cytometric analysis of leukocyte subsets: Simultaneous identification of five cell subsets using two-color immunofluorescence. Proc. Natl. Acad. Sci. 83:8361-8363.

Fig. 7.4. Reproduced with permission from Alberts B, et al. (1989). Molecular Biology of the Cell, 2nd ed. New York: Garland Publishing.

Fig. 7.5. Reproduced from Gray JW, et al. (1990). Quantitative cell-cycle analysis. In Melamed MR, et al. (eds). Flow Cytometry and Sorting. New York: Wiley-Liss, pp 445–467. The work was done at the University of California Lawrence Livermore National Laboratory under the auspices of the U.S. Department of Energy.

Fig. 7.8. Reproduced with permission from Dean PN (1987). Data analysis in cell kinetics. In JW Gray, Z Darzynkiewicz (ed). Techniques in Cell Cycle Analysis. Clifton, NJ: Humana Press, pp 207–253 (A,D); and From Dean PN (1985). Methods of data analysis in flow cytometry. In Van Dilla MA, et al. (eds). Flow Cytometry: Instrumentation and Data Analysis. London: Academic Press, pp 195–221 (B,C).

Fig. 7.10. Reproduced with permission from McNally NJ, Wilson GD (1990). Measurement of tumour cell kinetics by the bromodeoxyuridine method. In Ormerod MG (ed). Flow Cytometry: A Practical Approach. Oxford: IRL, pp 87–104.

Fig. 7.11. Reproduced from Darzynkiewicz Z, Traganos F (1990). Multiparameter flow cytometry studies of the cell cycle. In Melamed MR, et al. (eds). Flow Cytometry and Sorting. New York: Wiley-Liss, pp 469–501.

Fig. 7.12. Reproduced from Darzynkiewicz Z (1990). Probing nuclear chromatin by flow cytometry. In Melamed MR, et al. (eds). Flow Cytometry and Sorting. New York: Wiley-Liss, pp 315–340.

Fig. 7.14. Reproduced from Cram LS, et al. (1988). Overview of flow cytogenetics for clinical applications. Cytometry [Suppl] 3: 95–100.

Fig. 7.15. Reproduced from Gray JW, Cram LS (1990). Flow karyotyping and chromosome sorting. In Melamed MR, et al. (eds). Flow Cytometry and Sorting. New York: Wiley-Liss, pp 503–529.

Fig. 7.16. Reproduced from Gray JW, Cram LS (1990). Flow karyotyping and chromosome sorting. In Melamed MR, et al. (eds). Flow Cytometry and Sorting. New York: Wiley-Liss, pp 503–529.

Fig. 7.17. Reproduced with permission from Peeters JCH, et al. (1989). Optical Plankton Analyser. Cytometry 10:522-528.

Fig. 7.18. Reproduced from Gray JW, Cram LS (1990). Flow karyotyping and chromosome sorting. In Melamed MR, et al. (eds). Flow Cytometry and Sorting. New York: Wiley-Liss, pp 503–529.

Fig. 7.19. Reproduced from Gray JW, Cram LS (1990). Flow karyotyping and chromosome sorting. In Melamed MR, et al. (eds). Flow Cytometry and Sorting. New York: Wiley-Liss, pp 503–529.

Fig. 7.20. Reproduced from Van Dilla M, et al. (1990). Applications of flow cytometry and sorting to molecular genetics. In Melamed MR, et al. (eds). Flow Cytometry and Sorting. New York: Wiley-Liss, pp 562-604.

Fig. 8.2. Modified with permission from the Dako Handbook on Flow Cytometry, Glostrup,Denmark.

Fig. 8.4. Reproduced with permission from Yuan J, Hennessy C, et al. (1991). Node negative breast cancer: the prognostic value of DNA ploidy for long-term survival. Br J Surg 78:844-848.

Fig. 9.2. Reproduced with permission from Rijkers G, et al. (1989). Calcium mobilization in B-lymphocytes measured by flow cytometry. In Progress in Cytometry. Becton Dickinson, pages 63-73.

Fig. 9.4. Reproduced with permission of the Minister of Supply and Services, Canada, 1991, from Chisholm S, et al. (1986). The individual cell in phytoplankton ecology. Can Bull Fish Aqua Sci 214:343-369.

Fig. 9.5. Reproduced with permission of the Minister of Supply and Services, Canada, 1991, from Chisholm, S, et al. (1986). The individual cell in phytoplankton ecology. Can Bull Fish Aquat Sci 214:343-369.

Fig. 9.7. Photograph reproduced courtesy of Jan YN from Bier E, et al. (1989): Searching for pattern and mutation in *Drosophila* genome with a P- *lac* Z vector. Genes Dev. 3:1273-1287.

Fig. 9.8. Reproduced with permission from Krasnow M, et al. (1991). Whole animal cell sorting of *Drosophila* embryos. Science 251:81-85. © 1991 by the AAAS.

Fig. 9.9. Reproduced with permission from Krasnow M, et al. (1991). Whole animal cell sorting of *Drosophila* embryos. Science 251:81-85. © 1991 by the AAAS.

Fig. 9.10. Reproduced with permission from Weaver JC (1990). Sampling: A critical problem in biosensing. Med. Biol. Eng. Comput. 28:B3-B9.

Fig. 9.11. Reproduced with permission from Weaver JC, et al. (1991). Rapid clonal growth measurements at the single cell level: Gel microdroplets and flow cytometry. Bio/Technology 9: 873.

# Index

In this index, page numbers followed by the letter "f" designate figures; page numbers followed by "t" designate tables. Page numbers set in **boldface** designate glossary definitions.